Pilot Training

Pilot Training

Arthur J. Sabin

You Can Learn To Fly

WORLD

World Publications, Inc.
Mountain View, California

Library of Congress Cataloging in Publications Data

Sabin, Arthur J 1930-
   Pilot training: you can learn to fly.

   1. Private flying. I. Title.
TL721.4.S2   629.132'5217   78-24387
ISBN 0-89037-171-7 pbk.

World Publications, Inc.
Mountain View, CA

Dedicated to everyone—young or old—who has yearned to fly.

Contents

Introduction

What would motivate a forty-six-year-old professor of history and law, in the midst of a successful career as a teacher, writer, and law practitioner to climb into a 1,600 pound three-wheeled device and entrust his life to a contraption designed to fly? New horizons, maybe, revolt against the mid-life crisis, perhaps, just a yen to join an ever-increasing group of people who believe that in the coming years being airborne will be "the only way to go."

More than ten thousand people a month of every age, race, sex, and profession, are serious enough about learning to fly to try a demonstration ride or take an initial piloting lesson. The lure of flight is as irresistible now as it has been ever since the development of the first flying machines. This is true despite the fact that so many wouldn't voluntarily be a passenger in a light plane, much less undertake learning how to operate it.

This book is the story of one man acting out his boyhood dream of becoming a pilot. The dream started with enthusiasm that I fully expected to carry me along. Very soon, however, came the mature realization that learning to fly was anything but "kid stuff" in its demands and standards of proficiency. Along the way, from the first tentative questions about where to go for flight instruction, through gaining an understanding of the necessary skills, I found precious little in the way of guidance for the would-be aviator. Furthermore, there were areas everyone undertaking flight training should know about, such as the all-important safety factor, the costs in time and money, and the physical and personality attributes necessary. What really is involved in the learning process? How long does it take? Are special skills required? What about age and sex; do they make a difference? What tests do you take? Where do you get flight instruction? Just how safe is flying in a light airplane?

This book is designed to fill the information gap, but do not expect a glowing account of the "glories" of learning to fly. There is a great deal of fun, excitement, and personal fulfillment as well as pragmatic reasons for becoming a pilot, but there are stiff demands on students in addition to the limitations imposed on your ability once you become a licensed pilot. There are enough limitations to justify the basic statement of this book:

> Learning to fly is neither possible nor the right decision for everybody; becoming a competent pilot is difficult and most who try give up on the attempt.

Learning to fly is one of the great undertakings of life. In becoming a pilot you open up the potential of a dimension of living man has dreamed of throughout all history. In one sense you can enter the realm of space; moving, as man has always sought to move, as the birds—free of so many earthbound limitations. To fly is to be and become different

The choice, the decision to learn to fly, must be a rational and knowledgeable one. The decision must be made with some realistic understanding and appreciation of what goes into flight training and private plane piloting. For people who love to fly, the true picture of what is involved in learning how, the continued demands after you have a license in addition to accepting the limitations of the license, neither diminish nor destroy that love, but only render it real, stripped of absurd romantics.

In this book you will find one man's story of how he responded to a compelling need. I nurtured my early interest in airplanes through a good deal of experience in building and operating radio-controlled model aircraft. I do not necessarily recommend this kind of start for everyone, but it certainly gave me some valuable early preparation in the principles of aerodynamics and flight. Through the telling, within the framework of the current flight training scene, I hope the reader will better understand what is involved in becoming a private pilot today and better able to make the kind of intelligent decision about flying the undertaking deserves.

1

Making the Transition

Transitions are a part of our growing up process. The feeling of being at "the top of the heap" is part of us as early as our grammar school graduation. You are "big time" until a few months later when you find yourself at the bottom as a junior-high or high-school freshman—fresh, confused, and bewildered. The cycle continues, as you go to the top once again as a senior, and *make* the transition to the bottom as a college freshman or as a trainee on your first job. There is no denying that some cope with transitions with little reflection, perhaps little interest in the process itself. Others seem terribly intimidated by this transition.

The transitions of life become more difficult in some respect with the passage of years. Get past 30 or 35 and this transition process in some respects becomes more difficult to cope with, more demanding, and onerous in many ways.

Thus it was at this middle-aged period that I made a small decision; I went to visit a well-equipped hobby shop. That small decision involved a certain rational element, namely, I was so very deeply involved in earning a living with my brain that I felt the need for some alternative interest, a release from mental tension.

As a child, I had never been involved in sports to any degree. The dream of pro football or major league baseball was simply not part of my childhood ambitions. But I did enjoy working with my hands, building, taking apart, and fixing with some success. My parents thought I might be an auto mechanic because of my love of cars and my attempts to learn about their operation and repair. Instead, I ended up as a university and law professor. I looked in the window of Stanton's Hobby Shop thinking about the marvelous models on display and wondered whether I, a rank beginner—worse—not even having begun, could make the transition and find a level of success making models as a hobby.

I really wasn't sure what the undertaking would be; as a boy I had not too successfully worked with balsa wood and making the little wood kits that made those neat looking airplanes. In fact, I doubt whether I had ever really successfully completed any such kit.

I just wasn't good at building nor patient enough with it. In today's schools they would call it a "perceptual handicap." I found great difficulty following plans or "seeing" the next step in the building process and being able to translate it into actualizing a completed model. But I loved balsa wood, the smell of glue, and the idea of putting together a model. So, as I stared in Stanton's window admiring some of the larger radio-controlled models, I wondered whether I could cope with the transition to the bottom of a learning cycle, expecially when, considering past experiences, I might not be successful.

The concept of radio-controlled models held a certain mystery and intrigue. This was not "kid's stuff." I knew that because I had already read two or three magazines and found the language, the format, and the ideas directed toward a sophisticated audience of knowledgeable model builders.

Like most of these magazines, some attempt was made to keep speaking to the beginner but realistically, no magazine continues to talk to the beginner because it would essentially be reprinting much of the same material month after month. As I got deeper into radio-controlled modeling I found that the phenomenon of "talking to each other" had taken over. Manufacturers describe their products in terms that only the initiated would understand; the articles describing building techniques, illustrating suggested models to build, and describing the various sporting events (races, competitions, exhibitions, and such) all spoke a level of sophistication and in-group understanding that was both bewildering and, for the most part, quite incomprehensible. I knew the magazines were showing the "pros" in action, whether building, flying, or exhibiting their models.

I tried to make the transition in cautious, steplike fashion before making the real plunge, the purchase of a radio-control system, involving an investment of about $300. Underlying all of this was a basic question—was I ready to invest the time, effort, and money to becoming a radio-controlled-airplane pilot?

I walked out of Stanton's with a small model utilizing balsa wood construction and a modest-sized engine that could have a very simple, small radio-control unit mounted into it. Shortly after opening the box at home, I was back at Stanton's because I thought they sold me a kit with a major piece, the leading edge of the wings, missing. The clerk simply snapped the piece in half and showed me that the leading edge for both wings had been manufactured to come out as one piece of wood to be broken apart for two leading edges. Certainly not a very auspicious beginning!

I built that plane with some reasonable degree of success. The major thing was that I actually found my interest sustained to the point of completing the plane. From there I was back at Stanton's to purchase a model of an airplane I had loved as a child, the Piper J-3 Cub, the familiar yellow plane of years past. I remembered that as a teenager I had run off to a local airfield and plunked down two dollars (without my parents' knowledge) for a ride in one.

This second attempt at building was again reasonably successful though the time involved in construction began to give me some inkling that the commitment to building these models was of major proportions. On the other hand, there was real satisfaction in completing a model and being involved in an activity that was a far cry from the academic world where I spent my days.

The big plunge was to come soon thereafter. I had built two planes, neither of which were actually equipped with a radio-control system. If I wanted to go any further, it would have to be into the intriguing world of radio-controlled modeling necessitating the purchase of a larger sized radio-controlled airplane and a radio system.

Having studied the magazines with great diligence, I decided a trainer type airplane would be best for me. That meant a

high-winged plane with "full house" controls. A full-house
control means the radio transmitter would in turn control the
throttle, rudder, elevator, and ailerons of the model. I made the
decision to go full house rather than go the cheaper, simpler
route of a two or three-channel operation instead of this four-
channel "full house" construction. A "full house" is the easiest
plane to fly because of its ability to control all the major flying
surfaces.

The radio I chose was made by the largest manufacturer of
radio controlled systems in the United States, Kraft Systems,
Inc. These people put out a really fine product and stand
behind it with excellent service.

The building process was not an easy one as I attempted to
build what is called an "RCM Trainer" which stood for "radio
control modeler" and had some tie-in with the leading
magazine, *R/C Modeler*. It was a basic trainer, using a
reasonably small engine, and a tail-dragger (no nose wheel) but
otherwise used aileron control in addition to the other three
channels for operation. The kit is manufactured by Bridi
Enterprises, Inc., headed by a man named Joe Bridi (as I was to
learn). This was his first commercial kit and a quite successful
one. At one crucial point, it wasn't too successful for me
because I simply could not understand how to build those wing
tips, no matter how long I studied the plan and tried to figure it
out. Fortunately, I had by that time made the necessary
"connection," having been introduced through the mutual
friend, an advanced radio-controlled modeler Herb Meril. Herb
helped me over the parts I just didn't understand including those
wing tips and the installation of the radio.

How I admired his working techniques! He was neat, precise,
and knowledgeable. All the attributes I didn't have he seemed
to have, knowing exactly what to do, how to do it, and with
the precision and accuracy that left me in awe of a true master.
It was only with his help that the model got completed.

But once complete, the real question was getting the plane
into the air and learning how to fly. Herb did not want to teach
me, because, I believe, he had never taught anyone else and
feared making an error with my plane that would smash my
investment (and perhaps my hopes). Therefore, he introduced
me to John Bishop, a great big, lumbering, good-natured, kind

man who was a master mechanic with United Airlines. John's entire life seemed to be devoted to things mechanical, especially the world of flying and building radio-controlled models. I have no idea of how many people John taught to fly models, but I, as one of them, was undoubtedly one of his slower students.

It was a rather raw November day when we arrived at the field set aside by the Cook County Forest Preserve District for flying radio-controlled models. John looked over the model with its three-foot wing span and decided it was "flyable"; indeed, we got it running and off into the air without problems. A few adjustments (a process called trimming out) and John pronounced the model just fine for training purposes. In fact, to my amazement (and fright) he started doing loops and rolls with the plane that set my stomach into gyration and the blood pulsating with excitement and fear. I could not believe any model I built could do such things, and I half feared that even if it could, with my building skills, it was bound to come apart. But it didn't—at least not that time and for some months to come.

The learning process began in earnest the next spring and for months and months thereafter I would wait at the field, hoping John would show up to give me a flying lesson. He was very good about meeting me and offering help and I really stuck to it. He liked my enthusiasm, if not my skill.

Within a few months, a transition had taken place. Now there were people who were even more rank beginners than me, though I certainly was envious of those who could "solo" their training plane and who moved on to more sophisticated, sleeker, and faster airplanes, including models of World War I and II planes.

But I was learning, though at a painfully slow rate. I was slowed down even more when this first model came apart in the air (the wing folded in the center) and I had to quickly put together another.

By this time, another person had entered my flying life, Sid Axelrod, one of the owners of Top-Flite Models, Inc., a company that manufactured radio-controlled model kits (and other models as well). Top-Flite also is the world's largest manufacturer of model propellers in addition to marketing Super Monokote, a plastic covering for models.

I had written a letter to Top-Flite after building my second model (the J-3 Cub) and attempting to cover it with Super Monokote. There was no adhesive on the back of certain edges of the Super Monokote and yet there was no instruction or indication that this was supposed to be. To my surprise, Sid Axelrod called me, telling me he lived not far from me and would be pleased to give me a new roll to help finish the job. I declined his gift, but when he suggested that I come and visit him at his plant, that I could not resist.

It was from Sid that I got a second kit to replace that smashed "RCM Basic Trainer" and along with the kit he provided a good deal of encouragement. But I was still a long way from becoming a competent R/C pilot. Month after month I practiced with the aid of John Bishop and others. What I learned was that this was no "kids' game." The average R/C pilot is well over 30 (at least at our field) and the reasons became evident. It's not just the money, though that is a major factor; it is basically the requirement of coping with the transition as an adult from being a competent person in the field a person comes from to being a total incompetent, a novice, a rank amateur, and frequently a "performing idiot" in terms of building and especially flying radio-controlled models. The level of frustration, of defeat, and the need to take severe disappointment is such that most youngsters simply cannot handle; all are aspects of radio-controlled modeling. That isn't to say that many mature adults handled the matter well; coping with crashes was not an easy experience for anyone. You put love, time, devotion, and pride into a model and with one "pilot error," often involving a split-second of judgment, the model would become twisted into a crashed mess of balsa wood and destroyed hopes. Gravity is the most unforgiving force of nature.

The difficulties in radio-controlled modeling are immense, particularly in building competency in flying. It takes a great deal of coordination with respect to operating the sticks on the transmitter and visual perspective as you judge the speed, altitude, and direction of the plane in the air.

Finally the transition point took place; I had soloed and what a great feeling it was! John Bishop had prepared a small set of wings he pinned on me with that first successful solo landing.

The landing process is the most difficult in the radio-control model flying procedure and once you've soloed, it is a major breakthrough in the learning plateau. Two days after that, I completely smashed a model on another solo flight. Five years later, with continued involvement in the sport, I had built a number of planes and came to realize that ultimately I wasn't a very good builder at all. The process led me to reasonable competency in flying most kinds of radio-controlled models. I could fly low wing, high wing, biplanes, and planes modeled after World War II airplanes such as the P-51 Mustang and the P-47 Thunderbolt. Mistakes were still made, equipment failure sometimes meant the death of a model, but the learning continued. That learning experience had, however, taught me that there was a very real limit to how far I would go as an R/C pilot. I just wasn't that good in terms of becoming real "competition material," nor did I have the time to spend at the field to develop greater dexterity and accuracy, attributes needed to become a contest-type pilot.

Instead, I turned my attention to an entirely different area while continuing to do what could be called "sport flying." I combined my knowledge of law and my particular interest in the field of product liability with writing abilities to work for various companies in the radio-controlled modeling field. I also began to write for modeling magazines. The result was that I reached a kind of pinnacle, a kind of graduation point in that I was a recognized writer for the leading radio-control modeling magazine in the country, was writing two columns for magazines devoted to the modeling trade and working with model manufacturers on their product liability problems. I was being called into consultation by companies throughout the country to write their manuals of instruction, to advise them on their product liability exposures, and, for the most part, to help them communicate to the novice or beginner how to build and operate radio-controlled devices. I suppose one of the truly absurd facts of life is that I was able to write manuals on how to install and operate radio-controlled equipment and did a good job in that writing; yet I couldn't do a first-class job of installing radio-controlled equipment in a model or flying with top proficiency. I was the classic example of being able to tell others how to do it but not being able to do it yourself!

The transition process was at work again. My interest in radio-control modeling continued, with special interest and emphasis on writing for magazines and working for manufacturers. But something else was to happen in terms of transition; I glanced with longing toward the full-sized airplanes and wondered whether I could make the attempt to become a pilot of a full-sized aircraft.

It is, however, entirely unfair and inappropriate to talk about the transition from radio-controlled models to full-scale aircraft without some in-depth feeling for the sport of building and flying radio-controlled models.

For the uninitiated, it is often easier to explain radio-controlled modeling in terms of what is is not rather than what it is. The reference here comes from the realization that a high level of skill and dedication are involved in the sport. No one I know of was born with the skill to fly radio-controlled models. It takes dedication, a genuine commitment to undertaking the acquisition of those skills required to build a flyable model and, even more so, to acquire adequate flying skill.

Just to get a better dimension of what is involved, a few facts should amplify and explain the matter to provide a better perspective. The average radio-controlled model of the trainer type will be very similar to a high winged full-scale aircraft. One of the most famous trainer-type radio-controlled models is a scaled down version of the Cessna Skylane aircraft. This is because the flight characteristics you want in a radio-controlled training model are exactly what you want in a full-scale aircraft. The model and full-scale craft both will be "forgiving," meaning that it will not be so responsive or speedy that the learner will make fatal mistakes—fatal to the model or to the plane and pilot in the case of full-scale aircraft.

This trainer-type radio-controlled model will usually have a wing span of between four and five feet and about a four-foot-long fuselage. The power comes from an internal combustion engine obtainable in various sizes depending on the size of the aircraft to be used. These engines are incredibly powerful and require skill in properly controlling their power. They also co :titute a real danger if mishandled or abused. Even the most careful modeler will not escape having a propeller give his hand a good whack.

Most of the radio-controlled models now being built, even by beginners are full-house models, meaning, once again, that the radio control system operates at least four aspects of the model: the elevator, the rudder (often linked to a steerable nose wheel), ailerons, and throttle. The model on the ground and in the air can be controlled precisely in the same fashion as a full-scale model. The pilot has control over the three axes around which a flying model or full-scale craft move: the pitch axis operated by the elevator, the roll axis operated by the ailerons, and the yaw axis operated by the rudder with power from idle to full throttle operated by the throttle control linked to the carburetor.

The installation of the radio gear in the operating model is a very important skill. It involves careful placement, connection, and the installation of the servos to the moving surfaces (a servo is a small motorized mechanism used to actually move a part of the plane). A radio-control transmitter is a highly sophisticated electronic device, again, far from a toy. Using the very latest in electronic technology including integrated circuits, the trans-

Instrument panel

mitter sends out encoded signals through the air to the model. The model has a receiver or decoder that takes the signals and separates them, sending the appropriate signal to the appropriate servo. The servos are those marvelous little mechanisms that translate the electrical signal into a mechanical action moving the particular surface to which it is connected.

The radio-control system functions to translate the most minute movements on the transmitter "sticks" to the actual movement of the plane's moving surfaces allowing an astounding degree of control. Thus, once the engine is started and the plane is placed on the ground ready for takeoff, the throttle stick can be advanced in slow increments to full power and then, at the appropriate moment, the elevator stick can be moved just enough to get the plane into the air with the correct climb angle for takeoff.

From that point, the pilot has complete control over the movements of the plane in the sense that he can control its flight through all of the axis of a full-scale craft. He can roll it to the right or left, climb or glide, use the rudder to yaw the plane and fly at full or any desired power setting. When properly installed, the movements on the transmitter sticks are proportional to the movement of the surfaces so that the further the stick is moved in any particular direction, the further will that surface move. The result is a model that behaves with incredible accuracy as a flying machine, presenting a tremendous challenge to those willing to undertake learning of necessary piloting skills.

As those piloting skills are sharpened through hours of practice (and sometimes the loss of several models) the modeler will usually move to a low-winged craft that is faster, more responsive, and often more realistic appearing. Here the modeler can also get into modeling World War I and II airplanes. Usually these are not exact scale models because, for aerodynamic reasons, simply reducing a full-size craft to a five or six-foot-wingspan model will often fail to work as an operating model.

It takes real skill to learn how to fly any radio-controlled model with decent results. "Decent results" is being able to bring the plane down from the air in one piece. The process of learning is, for most, quite slow. For me, it was extremely slow in the sense that I was not particularly good at "seeing" what is required in terms of flying technique. For me, it seemed excruciatingly slow as I went through month after month of trying to build the necessary skills for decent flight.

Just as with full-scale aircraft, there is the necessary inspection of the plane before it is ready for takeoff. This includes checking out the radio, moving all the surfaces with the radio on and, for the wise modeler, having an experienced pilot review the plane to make certain the builder-pilot has installed the gear correctly, that the surfaces move in the right direction and that the plane seems to be flyable. The moment of truth for any airplane has to be in the air because no matter how carefully it was built, there is simply no way of establishing for certain that it will fly well or, in fact, fly at all.

Once the engine is started, the plane is taxied off to a runway or a takeoff made from well trimmed grass. Full power is applied and at the proper moment just enough elevator is given to get the plane into the air. (This "moment" becomes a learned-sensed feeling.)

One of the most difficult aspects of radio-controlled piloting lies in the dimensions, angles, and possibilities available to a flying object as it moves through the air. Furthermore (and this really gets a beginner) the controls are *reversed* when the plane is flying toward you as opposed to flying away from you. It takes many hours to build sensitivity to this aspect of piloting before it is successfully accomplished. Such skills must be learned, but once learned they do not rely on ordered technique as much as simply "sensing" what to do. Mastery is only there when the response is automatic; once automatic, it practically never fails because of pilot error.

By the end of my second lesson on the full-scale Cessna, I asked my instructor to be honest with me; to tell me when I should give up because I simply will not learn how to fly a plane. His response was that flying is only learned when the challenge of landing the craft is either met or failed. In a very real sense, the same is true of radio-controlled aircraft. It doesn't take too long for a reasonable degree of control to be successfully developed by the student R/C pilot once the plane has been taken off by the instructor and slowed down during flight and the transmitter handed to the student. After a few hours of practice the average R/C novice will be able to control that plane without getting into trouble too often; when trouble does occur, he may have learned to hand the transmitter back to the instructor fast enough for the instructor to get him out of that trouble. It is, however, learning to land that eventually separates success from failure in radio-controlled flight. In R/C piloting, it takes a great deal of perspective to be able to "see" the proper alignment of the plane with the landing field, set up a correct glide and bring the plane in for a safe landing. By "safe landing" is meant one where the plane is at least undamaged and at best brought in with such smoothness that, if you were riding within the plane, you would barely feel the touchdown.

At this point, I was still wondering whether the thrill of that first "solo" flight, the first time I took that R/C model off by myself, flew it around, and successfully landed would be as thrilling as my first solo flight in the full-scale craft.

As in so many activities, you have to see it and experience it to understand it; so it is with radio-controlled modeling. You have to see these models in flight, especially in the hands of a capable and proficient pilot to appreciate the beauty involved. To watch that model rise off the ground, perform beautiful loops, rolls, Cuban-eights, and other fancy maneuvers is simply thrilling and challenging to those who would attempt to duplicate succeeding in this area of sport endeavor.

In radio-control modeling, Murphy's Law is always operative—if something can go wrong it will. To the novice, every time something goes wrong with the model it's something so very new and different, and frustrating! To the experienced R/C modeler, the more you participate in the sport the less chance there is that anything remains unseen or unexperienced. By comparison, one distinct difference is the incredible reliability of full-scale aircraft. Everything basically behaves the way it's supposed to behave. At worst, in full-scale aircraft, some of the instruments may be a little bit off, but the dramatic contrast in predictability and reliability between full-scale and R/C airplanes is appreciated only by one who has been involved in both.

In summary, a lot can be said about this process of transition. In both you start off as a beginner looking at other

Student and instructor

beginners and wondering how far along they are in their
learning process. You come out to the flying field (either model
or full-scale) and you see a wide variety of students. You wonder
if they are rank beginners like you or more advanced. You find
there is really no way of knowing for certain. Still you wonder
where you are in terms of your training program as opposed to
other modelers.

The presence of a great many more women in flying full-scale
aircraft is evident. In five years of flying radio-controlled
models, I saw a woman fly a plane just once. Wives frequently
accompany their husbands out to the model flying field but
women generally don't get into the sport.

Certainly the feeling of helplessness, confusion, and the sense
that there is so much to learn that you are overwhelmed by it
all, is true for both kinds of flying. Besides the common factor
of being "overwhelmed," both the R/C pilot and the student
pilot of full-scale aircraft have been lied to. The lies consist of
sales pitches that talk about the ease with which you can build
or fly a radio-controlled model or that "if you can drive a car,
you can fly an airplane." These are simply lies and distortions.
Success in either field takes a combination of certain basic
capabilities that you either have or don't have, along with a
great deal of perseverance and willingness to take punishment

in forcing a learning process. I say "punishment" because most human beings feel swamped by what is demanded of them as an R/C pilot or a student of full-scale craft. To those who take the matter seriously, to those who really want to succeed in the sense of becoming decently proficient, competent, and confident as pilots, there is "punishment" in store. The question is whether it's worth it or not, and only the individual involved can give that answer. Most people who undertake building radio-controlled models never succeed in flying them with any level of competency or success. I know from the "inside" that even after the substantial investment in the radio system ($200 to $600) the majority of radios never get flown in a plane. The few that are flown are only used once or twice with the help of an instructor, the novice pilot rarely gets beyond the first or second flight; when I asked my instructor what percentage of individuals undertaking the pilot training course for full-scale craft completed that course, he hesitated but came up with 75 percent completing the course and getting a private pilot's license. I didn't believe it, and have since found that only about 40 percent actually get their license. It isn't the cost usually, because roughly the cost of being in the R/C modeling hobby to the point where you are a proficient pilot can mean an investment comparable to the investment made to get a private pilot's license, roughly about $2,000.

The transition stage for me involved moving from the position of a reasonably competent R/C pilot who had developed an expertise not so much in flying as in understanding and writing about R/C modeling, to that bottom rung on the ladder of learning to fly full-scale planes. The beautiful advertisements, the alluring ads promoting learning to fly quickly change to the reality of a gruelling, demanding learning process.

Undoubtedly, the transition is easier for younger people and those who have not reached a level of proficiency so that they haven't been beginners at something important for a long period of time. In this respect, the transition is equally difficult for all, regardless of background. If it's been a long time since you had to contend with being at the bottom, with feeling inadequate and foolish, it will be more difficult. But most can contend and cope.

2

Seeking Pilot Training

When I called a place advertised as a Cessna Pilot Training Center in the telephone book and asked whether an appointment was necessary, I was told in a most casual manner, "Naw, just come on in." I started to ask some questions but it was apparent the young woman on the telephone didn't have the answers. This was not encouraging. If I was really going to make the move and see if I could cope with the problems of undertaking flight training, I needed all the encouragement I could get.

I had learned that need for encouragement in my role as instructor of novice R/C pilots. I discovered that beginners need encouragement and admonitions about being patient and not demanding too much of oneself. It's what I needed, it's what every beginner needs in undertaking any challenging endeavor. If I didn't get encouragement from the woman on the telephone, would I get it from the flight instructor?

Having reviewed what was available in terms of flight training, I learned that the options included:

- FAA-approved schools or non-approved (but regulated) schools
- FAA-certified flight instructors offering private lessons
- flying clubs

FAA-approved schools or non-approved (but regulated) schools. The approved school is one that will carry a flying school rating, a ground school rating, or both. They also may be authorized to give their graduates flight tests and written tests for the private pilot certificate. Schools that do not have the approved status may offer flight courses, but they cannot do everything an approved school can do. The approved schools are generally ones that require residence for training. This is the most efficient and thorough way to undertake pilot training, but it is not practical for the average person.

The non-approved, but regulated, flight schools are the ones advertised in your local newspapers and Yellow Pages. They are most frequently tied to an aircraft manufacturer. The tie-in is that they offer an FAA-approved course through their dealer network, built around flying their particular aircraft. Leading manufacturers, including Cessna, Piper, Beech, and Gulfstream American (formerly Grumman) have organized these pilot training courses.

By writing to the U.S. Department of Transportation (Publications Section, TAD-443.1, Washington, D.C. 20590), you can obtain FAA advisory circular 140-2J, "List of Certified Pilot Flight and Ground Schools." This brochure is available free of charge. The FAA General Aviation District Office can be contacted for a list of all active pilot training facilities in your area. They maintain a current file on all certified schools and instructors within their district. Look up the FAA in your phonebook.

FAA-certified flight instructors offering private lessons. Instead of signing up for a course of instruction as you would normally do under the previous options, you are paying by the hour for your dual instruction and getting the guidance that can be given by that flight instructor with respect to your ground school study.

Flying clubs. Flying clubs will sell you a membership or share in the club. In return, you get flight instruction from a CFI (certified flight instructor) and share time and expenses on the club's airplane or airplanes, at a much lower cost than individualized instruction.

The most popular and well organized of the three is the non-approved, but certified courses usually offered by aircraft dealers. These were started about ten years ago when the aircraft manufacturers began to realize that the lack of interest in flying was largely because of the poor quality of instruction. Potential pilots needed training on an organized, integrated basis where flight training and ground school were reasonably integrated and offered reasonable, meaningful, instruction. All major manufacturers of light, single-engine, fixed-wing aircraft now offer a course in private piloting training.

I had contacted the FAA for their list and noted the Cessna dealer and its Pilot Training Center at Midway Airport. I chose Midway because of its proximity to downtown Chicago, figuring (correctly as it turned out) that I could get back and forth during the middle of the day without too much traffic hassle. Midway, years ago an alive, busy airport, all but died except for small crafts, once O'Hare International Airport was built following World War II.

Once at Midway Airport, I found the Pilot Training Center. There was the young lady, quite disinterested as she had been on the telephone. The place was almost, but not quite, as bewildering as that first visit to the R/C field.

I asked to speak to someone about starting pilot instruction and was introduced to a slim young man who could not have been more than 25. Tom Murray answered my questions with simple, direct statements. He didn't try to sell me; he just listened and suggested the "demonstration ride" for $10. I decided to take the demonstration ride.

I had been in small aircraft before and, along with my wife, once had a delightful ride in a high-powered Cessna aircraft out of San Diego. The little Cessna 150 Tom and I climbed into was a very stripped-down, non-luxurious model of that larger Cessna.

Tom spent about an hour (the cheapest hour I was ever to spend in any aircraft) demonstrating the plane in terms of pre-flight checking, takeoff, flying around the field, and finally landing.

I had absolutely no sense of fright or nervousness, but I had not expected to. I knew what was going to happen and what it would feel like. My mind was not on the plane or the experience of flying, but rather on whether I was capable of learning the skill and artistry of flying.

When we returned to the office, I got only a brief explanation about what the pilot training course consisted of. I saw the "student room" where audio-visual materials were kept (there was no one in it) and discovered it would cost about $90 for "signing up for the course" which committed you only to buying the textbooks and allowed you the privilege of using their facilities for study. I took the obvious course—I decided to think it over. That's just what I did...a great deal.

3

Flight Training Begins

It's now some three weeks since the first formal lesson in flight training began. Obviously, my decision was to begin my piloting "career." Tom had strongly recommended at least two lessons per week and it seemed to make a good deal of sense. He explained that if you only flew once a week, too much time was spent in simply redoing that which had been done before to make any real progress. With this I heartily agree. It is all too evident that there is so much to learn and such an important need for reenforcing that which has been partially learned that involvement in the process of learning how to fly requires conserted effort in putting aside the time on a very consistent basis. Unfortunately, that "consistency" is so often interrupted by weather; you simply have to have the cooperation of the weather in order to have those flight lessons.

Without reasonable clear visibility, unless you are at the point where you are trying to accomplish some advanced training in instrument flights, it is both senseless and dangerous to fly in anything less than very good weather. This is emphatically true for training as well as for flying after achieving your Private Pilot's license. Thus the student has to face the disappointment of having flight sessions cancelled, often with excruciating regularity; the weather always seems to turn bad just when you have a lesson scheduled.

The flight lessons themselves consist of a block of two hours of time. For that two hours of time, you will generally be flying about 1 to 1.3 hours; the rest of the time will be devoted to getting to the airplane from the office where you meet your flight instructor, discussing what the lesson will be covering, and doing the necessary run-through on checking out the plane before climbing in. Then there's the necessary checklist for starting up the plane, followed by getting the craft out to the proper runway and doing the runup and checking before takeoff, all of which eats into that two hours.

Additionally, that two hours is going to include the time it takes to get back to field, getting into the landing pattern, land the plane, taxi back to its parking spot, and going through the checklist before shutting down the engine and finally tying down the plane. Then it's back to the office to enter the flight into the logbook that every pilot, student or licensed, keeps with respect to every hour of flight and, let's not forget, that great moment—paying the bill for the lesson.

As in almost all facets of our economy, "plastic money" is quite acceptable in paying for flying lessons. You are not usually committed to the completion of a course, therefore you "pay as you go" which allows you to quit if and when you decide to do so. As of this writing, I think a realistic figure for the number of hours of flight training and ground school preparatory to taking a private pilot's license examinations would be in the neighborhood of $2,000. Remember that you're paying by the tenth of the hour for all flight instruction; they have a nice little meter that clicks off exactly how much time has been spent with the engine running. In addition, you pay for your instructor's time including when the CFI works with you over your periodic ground examinations covering flight theory. Yet for those seriously involved in learning to fly, I for one would rather have the structure including the business formality of this kind of training as against a "friend" (who in any event must be a certified flight instructor) who will "teach you to fly." With as many misgivings as I already have with respect to the training received, it seems to me to be a far better system to know that you are paying for the service of instruction than depending on "friendship." These certified flight instructors are just that; they have a certain degree of expertise in training of those who wish to fly and frankly I am simply more comfortable (as I think most people would be) with the business relationship involved. Learning to fly is also a business undertaking, not just a "lark." If nothing else, it emphasizes the fact that you are making a serious investment of your time and money and not simply experimenting with some fantasy.

The fantasies about learning to fly disappear quite rapidly once the individual is actually emersed in the process of learning.

That process of learning is, by the way, much more than that two-hour flight session; it involves the process of an intensive course of study—book learning for the want of a better means of expressing it. For example, when you undertake the Cessna Integrated Flight Training System, a great deal of that "integration" involves going through a very heavy, long, large book called the *Manual of Flight.* I would estimate that for the average individual, for every hour of flight training, three hours must be spent on the other learning aspects involved in flight training and pilot preparation. Along with the flight training manual comes a workbook and you are immediately faced with the fact that this isn't kid's stuff in terms of the questions asked; most of them are really tough, and they should be, because becoming a proficient pilot is a tough and serious undertaking.

Some assistance is provided by an audio-visual system involving a cassette player with a tape and a book of pictures which the student uses, preferably before the section of the flight manual to be studied. The literature put out by Cessna indicates quite clearly that the audio-visual materials are available for home study. This certainly was not true of my experience; a room was set aside at the Pilot Training Center for the audio-visual work but none of it could be removed. Therefore, for the conscientious pilot (and I cannot imagine anyone really becoming involved in flying without being conscientious about the matter) a great many hours must be expended in just plain, hard study. You've got to hit those books and hit them hard. These studies are not only to explain the theory of flight and, in effect, to explain in writing what you are attempting to accomplish in flying, but also constitutes the necessary preparatory work for the written part of the private pilot's examination process; the FAA requires that you successfully complete a written examination as well as a flying examination.

And is there a great deal to learn! Even making this transition out of radio-controlled modeling where the difference between an aileron and a rudder is second nature and where aspects of flight are well known in terms of the flying behavior of models, the point that has to be made is that you are entering an entirely new discipline. That discipline is a demanding one, involving emersion in such facets as the basic physics involved in flight, an understanding of basic weather phenomena as well as acquiring an understanding of the entire world of flying. By "world of flying" I mean everything from getting acquainted with Federal Aviation Regulations (FAR) pertaining to flight, the use of radio transmissions, the operation of the power plant (engine) of the plane, and the computing of the necessary and vital statistics of flying—weights, rates of ascent and descent, distances travelled, fuel consumption, etcetera. Then there is the entire matter of knowing where you are going, involving the study of navigation. On this latter point, it strikes you after a few hours of flying instructions that this might all be great, but without the help of that instructor, how could *you* ever find your way back to the airport? This is no simple process, and I am certain that there are individuals who can and do learn to fly in terms of controlling and being in control of an aircraft and yet simply cannot navigate from one point to another. One of those people might be me since I have such a poor sense of direction.

I have earlier referred to certain fantasies about flying, more exactly about learning to fly. It doesn't take many hours of actual flight instruction and study before the realities come into perspective. Interestingly, I must have read at least four books by individuals who had learned to fly and who were essentailly exuding the thrills, pleasures, and benefits without really talking about some of these realities and hard facts; hard in the sense that they are real and, if understood, do not necessarily destroy the desire to learn to fly, but certainly are part of the reality of the process.

Perhaps one of the first realities is that the third dimension— the dimension of flight—is a hard dimension. We're simply not used to it, and understanding it is complex both from a theoretical standpoint as well as from the pragmatics of learning to behave as an individual in that third dimension and control a machine in that dimension. Don't let anyone kid you; the fact that the aircraft can gain and lose altitude is the salient difference between that and any other human activity and it makes for one hell of a difference in its demands on the individual who would conquer that dimension.

Another very real fact is that while flying has become a lot safer than it was, say twenty-five years ago in terms of the quality of the aircraft and the instrumentation available to the pilot, it is also true that as flying becomes more technical it becomes more difficult. I recently had the opportunity of looking in at the cockpit of a World War I "Jenny" trainer, used to train American pilots going to France for war duty. The dashboard consisted of three or perhaps four instruments, and that was it. The maze of instruments facing the student pilot in even the most elementary training airplanes of today is bewildering and requires a great deal of acclamation as well as understanding. It is apparent that as one becomes a proficient pilot, that person becomes "sensitive" to those instruments though barely looking at them (and in fact at times, I'm certain, not looking at them at all). One learns to fly through the "feel" of the plane as well as what the instruments "should be saying" in terms of the aircraft's performance.

Now that I have begun that flight training, nothing is more ludicrous than to think about some of the themes developed and statements made in the books I read prior to undertaking flight training; the "theme" that flying really isn't very demanding and what must be termed the deliberate absence of any honest discussion addressing the fact that the person who would want to learn to fly is a person who must be willing to enter a new, complex, bewildering, and demanding world.

And how safe is that world? As indicated earlier, the average person's idea of safety is precisely wrong; they have the idea that it is the engine or some part of the aircraft that makes flying "unsafe" or potentially dangerous. The statistics that are available from United States government sources clearly indicate that it is not the aircraft that is unsafe, but the pilot. Roughly 90 percent of all accidents resulting in fatal injuries have nothing whatever to do with the aircraft; they are "pilot error" in the sense that the pilot flew into or got caught in adverse weather, that the pilot lacked sufficient training, or that the pilot simply made an error of one sort or another. Now there is no question that you can "play" with safety statistics; media also do interesting things to distort the scene. It is true, for example, that every time a small plane crashes and someone gets killed, it makes for "good reading" in the newspapers and "interesting" viewing on television. This exaggerates the real picture; consider the fact that in an average Memorial Day weekend, five hundred or more Americans will die on the highways. The number of people who will die in noncommercial, private-pilot flying for the entire year of 1977 (the latest available as of this date of writing) amounted to 1436, passengers as well as pilots.

On the other hand, there is very little margin for error in flying. That perhaps is the scariest aspect of learning how to fly. Yes, you can make a mistake on the road; you can run the stop light or change lanes without seeing that car in your blind spot and get into a fender-bender. While it is practically impossible to gather reasonable statistics on the number and kinds of automobile accidents, the plain reality is that a sizable margin for error does exist in the sense that you can bend fenders, smash up automobiles in a variety of circumstances and yet come out unscratched or with slight injuries. Flying, on the other hand, is unforgiving of even slight errors, and that is perhaps the big difference. That which balances that big difference is the level of training and the exacting demands of the government in certifying an individual as a private pilot, a person licensed to fly. If the standards required of a driver were comparable to what is required in flying, the accident rate in the United States would go down tremendously. Furthermore, the government has its ways of keeping your abilities and knowledge at a reasonable level in order to maintain your right to fly, a concept that is just beginning to come into the licensing requirements of some states for driving.

The safety of flying obviously involves a number of components; it begins with a safe aircraft, and, as indicated, the aircraft itself is most often going to be the safest aspect in the "safety question." To stay on that point for just a little while, there are no "alley mechanics" that can work on airplanes.

They all have to be rigidly trained, licensed, and certified. Furthermore, for an airplane to have the "right to fly" it must have a complete annual inspection in which every aspect of the aircraft is carefully gone over and all repairs must be made before the aircraft can be recertified for another year of flying. On the other hand, there have been enough accidents involving the alleged failure of some part of the aircraft so that there are now pending lawsuits with claims against the manufacturers of light aircraft which, if liability were proven and the amount claimed paid, would wipe out every aircraft manufacturer in the United States many times over. As an attorney, my eye has stopped short from time to time as I have read about the decision in some case involving the accusation of a defect in design or construction of some aircraft resulting in some disaster. At least in terms of the allegations made against these companies, callousness concerning the safety of the plane in design and construction begins to approach that deliberate disconcern with safety so frequently reported concerning the design and construction of automobiles.

But it is the pilot and thus the human error that must be understood to be the major component of flight safety! In turn, pilot competency is not only a matter of actual in-flight training and the "book learning" aspect of that training, but also depends upon judgmental qualities as well as physical attributes of the individual. Judgmentally, there is the ultimate question of whether the pilot should even take off, given reported weather conditions. His particular physical condition or emotional state at that point are also crucial. Part of that safety component is also seen in the continuing judgmental process which has so many facets that it simply cannot be adequately articulated; should you "go around" rather than complete the landing; have you understood the instructions of the tower control; are you going to land on the right runway and in the right direction; are you capable of the precise, often emergency-type reactions to flying situations? All of these go into flight safety and in a very real sense constitute both every tangible as well as intangible factors in assessing just how safe you think flight is for you as an individual as well as in terms of the general picture of the safety of flying itself.

It is kind of a half joke, but unfortunately true, that there is still a great deal of argument as to exactly what makes an air-plane fly. Given the perspective that heavier-than-air flight by man has progressed for less than one hundred years since its inception with the Wright Borthers and their twelve-second flight in 1903, this lack of firm knowledge is not so astounding. On the other hand, man has flown to the moon and flies at supersonic speeds; yet the question remains and is argued. Furthermore, as a student pilot you are going to come across differences in opinion about what to expect out of that airplane. For example, I quote from a book written by Wolfgang Lingewiesche, called *Stick and Rudder,* an explanation of the art of flying, which is considered a classic in its field though written in 1944, a time in the development of flying that some consider the "Middle Ages" of flying. He has a chapter called "What the Airplane Wants To Do" and says,

> That the average airplane really *wants* to spiral dive is the cause of that ef-fect that has cost many a life and should be more clearly appreciated by many people, especially by the people on the "outskirts" of aviation—stu-dent pilots, teachers of aeronautics in high schools, aviation "officials" of various sorts. The exact causation is too involved to trace here, but the effect is this: *if an untutored person tries to fly an airplane* and uses the controls in the manner that seems most "natural" to him, responding most energetical-ly and most quickly to those disturbances that "naturally" impinge most sharply on his consciousness, *the flight will almost certainly end in a spiral dive and a crash.*

Now compare that to this paragraph coming out of the Ces-sna *Manual of Flight* as it talks about one aspect of the stability of the airplane,

> Longitudinal stability is pitch stability, or stability about the lateral axis. Most training airplanes have good longitudinal stability, which means that they will stay at a level flight attitude rather well without con-stant attention from the pilot. They will tend to return to level flight if the pitch attitude is disturbed by rough air.

Given that the two authors are not talking about exactly the same subject and admitting that the latter book does deal with spiral instability in another portion, there is no question but that you are getting two impressions about what the airplane "wants to do" in terms of its flying characteristics and the ability of the untrained individual to control or respond to what the plane "wants to do."

In summation on these points, the realities that bespeak learning to fly involve the readiness to enter and conquer a new world. That it is exciting and challenging cannot be denied. That it is worth the price in terms of its demands can only be answered by each individual who undertakes to conquer. The disservice that is done by so much of the literature is to make it sound so simple, so readily attainable, effectually avoiding the components of the *real* demands made upon the individual who wants to become a pilot. Yes, attitude is important in terms of learning the skills through in-flight training as well as "hitting the books." But that attitude has to include the component of perseverance and the willingness to learn a great deal very quickly without getting discouraged as errors are made in the seemingly overwhelming task of understanding this new world is impressed upon the student pilot. You are also entering a risky area of human endeavor, one where mistakes are not "forgiven" and can end in the loss of life; anyone who believes that only happens to "others" is a fool. What in incredible is the number of the "you'll love to fly" books that never discuss in hard terms the danger in flying and the razor-thin margin for error.

Furthermore, don't let anyone, including licensed pilots, much less the literature put out by the aircraft companies, kid you; the private pilot who is already licensed has often forgotten all that he had to go through because he's gone through it. The business aspects and advertising nonsense put out by the aircraft companies speak for themselves. What's real is real; it's tough, it's demanding, and it takes a good deal of time, deep concentration, exhaustive effort, and ultimately you may learn that you haven't got what it takes in terms of the physical or mental capabilities (and this has nothing to do with intelligence) demanded of one who would become a pilot.

Competence is not just a "nice to have" attribute—it's crucial in the need to survive to fly again. Like so many activities, you don't know whether you will succeed until you try it; you don't know whether you will love it until you are ready to say to yourself that you are willing to pay the price to get there. Interestingly, on a much different (though relevant) level, the same is true of learning to fly a radio-controlled aircraft.

Over a recent dinner table conversation, our daughter, about to embark on her university career as an entering freshman, expressed what might be termed the usual fears and anxieties but also the underlying excitement involved in this threshold experience. In my response, I not only gave the usual soothing reassurances (she was an excellent student in high school, had good study habits, etc.) but also remarked on the idea of relishing the position as a novice—a beginner—a freshman. I reminded her that she would remember these times with a feeling that they were, indeed, exciting times of new experiences and involve the pleasure of achieving reasonable success in meeting the challenges presented. In effect, I was asking her to realize that some day, perhaps not too soon from now, she might become almost nostalgic about that bewildering, confusing, demanding, and yet exciting "time of beginnings." What I was pleading for her to do was to appreciate the delights of precisely being in the situation of a beginner.

As the conversation ran on, I began to reflect upon my own words and ideas in terms of my experience as a beginning student pilot. Were these ideas and attitudes not similarly relevant to the experience of undertaking learning how to fly? I asked myself whether I could honestly relish being in the position of the beginner, and the answer that came back was that unless enough pleasure is found in the demands and the work of learning how to fly, then indeed the beginning would also be the end in terms of never having gone very far with the undertaking. Plainly put, the "mental set" must be such for the individual that there is at least some measure of real pleasure that cuts through the bewilderment, the confusion and the demands to make the learning process worthwhile.

Having logged some ten hours or so of actual flight instruction, more of the nonsense disappears and more of the realities become evident. I do not believe it can be overemphasized that learning how to fly is hard work, just plain, hard work. Spend an hour in the plane with the instructor, both monitoring and demanding performance, and you've got the equivalent of four or five hours of really hard work, both mental and physical.

It may be surprising that I include the word "physical" in

describing the work. No, you are not shoveling coal, but it is physical work and not simply in the sense of coordination. There is physical strain involved in holding the controls in certain rather rigid positions, especially when attempting to maintain a critical air speed. What you learn, that is, what you have to learn, is the crucial ability to make that plane fly slowly. Without that ability, landing the plane can never be appropriately mastered. It takes real physical effort as well as proper judgment to hold the controls in precisely the right position in order to obtain and then maintain that slow flight. Undoubtedly, experienced pilots know how to use the trim control or controls so as to alleviate or relieve a good deal of the physical strain but until you're able to do this automatically, you're going to find the muscles particularly in your left arm will be strained after your flying sessions. It's the left arm because that is the one holding the control wheel while your right hand is on that throttle. It is the combination of the precise control stick attitude (the right amount of back pressure) and the throttle setting that determines your ability to control the plane in slow flight.

While on the subject of slow flight, it should be understood that for a training airplane like the Cessna 152, the most common of the trainer-type airplanes, slow flight without flaps extended in order to yield a sharper angle of decline, is 65 knots. It is between 60 and 65 knots that you learn to land the airplane. Translated into miles per hour, that seems pretty fast, and it is except when you realize that the big jets are moving between 140 and 180 miles per hour when they're approaching the runway. That's another of those "facts" that you won't find generally known or understood.

Well, with these ten hours or so logged in my Pilot's Log, where do I stand in terms of skills? At this point, one becomes reasonably comfortable with the pre-flight check of the airplane. By "reasonably comfortable" is meant that you can do it right though you continue to use your checklists very carefully. One comforting aspect is that careful pilots continue to use those checklists regardless of how familiar they are with their aircraft. Once again, if automobile drivers had to perform the checklist procedure as rigidly as is required in flying, the number of accidents would drop significantly as well as a diminution in such embarrassing moments as running out of gasoline, trying to drive away on a flat tire, etcetra.

Furthermore, one begins to feel reasonably comfortable in terms of understanding what it takes to get that airplane into the air. After the pre-flight check in which the external components of the plane are visually tested and you climb in and go through your checklist to start up the engine. You become by this time reasonably proficient in going through the start-up procedure with respect to the engine and the elemental settings of the instruments. This includes setting the altimeter according to the barometric pressure given for that hour and then setting the heading indicator which is a gyro-compass used in your navigation of the plane. You will use this directional gyro-compass much more than the little magnetic compass hanging from the ceiling in the center of the windshield. You have also learned how to dial in the radio to the correct frequency in order to get the necessary information for the setting of the altimeter and certain basic weather information as well as which runways are to be used for takeoff and landings.

But before moving that plane one inch, you must begin initiation into that quite terrifying world (even for many experienced pilots) of speaking to ground and tower controllers. In most larger airports, you have a separate ground controller on a special frequency and then a different air controller on the air-control frequency. If you have ever seen anyone proficient at speed reading and noted with amazement their ability, then you

Cessna Model 152 II

Over water

must understand that ground and tower controllers are masters of speed speaking. You can (and must) listen to these people for hours in order to begin to understand with some confidence what it is that they are saying and in fact when they are speaking to you! So now at this point, after these hours of flight training, I am reasonably capable of notifying ground control where I am ("Cessna 757 Bravo Uniform, at the West ramp, taxiing with Zulu"). That tells the ground controller who you are and where you're going because you have listened to (on another channel) the information that has given you the altimeter setting, the weather, and the runways for takeoff and landing, with that information identified by, in this case, the word "Zulu."

Another aspect of the world of flying is to learn the international phonetic alphabet which begins with Alpha and ends with Zulu. When you have told the ground controller that you are taxiing with Zulu, he knows where you are going in terms of the runway for takeoff because he knows what the "Zulu" information was.

On the one hand it is comforting to know that English is the international language for piloting but that does impose the burden on the student to learn the international phonetic alphabet. I found that the easiest way to do so was to make use of the fact that in our state many of the license plates carry letters in addition to numbers. Thus, as I drive, I mentally recite

Over land

(sometimes out loud as well) the phonetic equivalent of the
license plate. I also practice as I am driving the necessary
language to speak to ground controllers and air traffic
controllers in order to say the right thing and say it quickly
enough.

Once ground control knows who you are and where you are
going, and you have an idea of what ramps to use in order to
get to the correct runway, you begin to learn the technique of
controlling the airplane through the use of rudder control.
What is essential is that you have to "unlearn" that which you
have learned from driving a car. A plane is steered by use of the
rudder pedals; the wheel in front of you has no control over the
direction of the plane on the ground. This learning and
unlearning process is not easy for one who has driven a car
throughout their adult life. There is that instinctive reaction to
try to steer with the wheel rather than with your feet. It is
further complicated (and that complication is already there by a
few lessons) by the fact that you have to turn the wheel in a
particular direction in order to move the ailerons in order to
compensate for cross winds pushing against the airplane on the
ground in order to avoid tipping over or other unhappy events.

Once you get near the runway for takeoff, you then have to

park the plane at the side near that runway and go through a "run-up" procedure prior to takeoff. There are brakes on an airplane that apply to the right and left wheels; they are found on the tips of the rudder pedals and are activated by pressing with your toes, so you see that some real coordination is necessary. The brakes are used mainly to accentuate the direction of turn rather than stopping the airplane itself. They are also used to hold the airplane in place while the engine is being "run-up" for test purposes.

At this point, I have become reasonably proficient in parking the plane near the takeoff runway and going through the procedure of final checks before takeoff. This includes running up the engine and checking the magnetos along with other procedures to make certain that all is functioning well, including the moving surfaces on the plane. Once you are certain that all is well, then you dial in the tower frequency and tell them that you are ready for takeoff and they in turn will come back with a quick response (it's always incredibly quick) telling you either to hold or that they are cleared for takeoff.

With a reasonably large airport and their reasonably large runways, taking off the plane is not a difficult procedure. All you have to do is learn that coordination of the rudder pedals while applying full power and then, when the correct air speed is reached, pulling back on the wheel which in turn will raise the nose of the plane off the ground and then your job is to keep the correct air speed and a proper rate and angle of climb as you leave land; you have now entered that third dimension—you are flying in space.

Part of these ten hours I have been involved in the usual training of taxi practice, takeoffs, and the procedure for climbing to a desired altitude. From there, the training begins with the concept of how the airplane should look in a straight and level flight in terms of your visual reference from the plane to the horizon and in terms of what the instruments tell you about the attitude of the plane. With all beginners, there is a tendency to concentrate too much on the instruments and what they relate as against what you should be learning from your visual reference to the horizon and the attitude of the nose and wings with respect to that horizon. You've got to learn to have one hand on the wheel, one hand on the throttle, the rudder pedals worked appropriately, all with the intent that you achieve proper straight and level flight in a given direction at the desired altitude. Once achieved, you learn to trim out the craft to hold that altitude and heading.

Once again there is a very significant difference from driving a car; with the exception of cruising in a set direction over a substantial distance, in flying, your right hand has to be constantly on that throttle; this physical position itself is really different from anything else you will experience or that would be demanded of you in driving.

In these early hours of flight training, the major lesson beyond getting the plane to head in a direction at a specific altitude and at the normal cruise speed of the plane, is the process or learning to turn the plane by banking its wings. Here is where you will use the wheel to the right for a right turn and to the left for a left turn and use the rudder pedals for coordinating that turn. Learning to sit straight and not "lean into" or "away from" a turn, the correct amount of aileron movement and back pressure on the wheel to compensate for the loss of lift in turning takes concentration and practice.

Once you get some feel for this aspect of flying, you begin immediately to learn the technique of slow flight which, as mentioned earlier, takes a great deal of coordination and must be absolutely mastered in order to land the plane.

Critical to all of this is the fact that the wheel, in terms of forward or backward pressure determines the air speed while the throttle, controlling engine speed, controls the altitude. This is precisely the opposite of what one would logically perceive from driving a car. And is this difficult to learn! In ten hours of flight time, I couldn't count how many times Tom has said, "Watch your air speed" or helped me by translating it into "Your nose is too high—your air speed is too low."

This whole business of slow flight, where you're slowing the plane down to an air speed of about 60 or 65 knots per hour and maintaining the same altitude (not going up or down) is also vital in terms of recognizing when you are getting into the position of stalling out the airplane. When an airplane stalls, it basically noses straight down even though you're pulling back on the wheel which would normally make it go up. In fact, when you reach that critical stall speed and position (an airplane can stall at any speed but usually a stall occurs at low speeds in terms of actual flight) pulling back on the wheel simply makes it stall faster and may result in a spin. Therefore you have to learn slow flight and then experience stalling out the plane so that you learn to recover from a stall. This involves recognizing when a stall is imminent and then either keeping the airplane from stalling or purposely having it go into a stall in order to learn how to recover from that stall. All of this is practiced (but it seems never enough) in those first ten or so hours of flight training. There is a stall warning horn that comes on with an eerie buzz letting you know that you are reaching the stall position; this development is a real advance for aircraft safety and flight operations and was developed, surprisingly, just a very few years ago.

As a personal aside, throughout this initial ten hours of flight instruction, I found Tom to be constantly demanding more of me than I really was ready to give. I would have preferred a great deal more review than was allowed or given. In fact, there is practically no review except in the sense that certain flight aspects will, in the normal course of a flight, reoccur; this is to be distinguished from a deliberate attempt to review that which has been learned. One incredible aspect of this is that when I look at the syllabus that is the heart of the Cessna flight training, it moves even faster than Tom moved in terms of pressing me to learn more and more. In a word, I have got more hours of flight time and have learned less than they expect in the syllabus! Now, does that make me a dummy? Does that mean I lack the coordination and ability to fly? According to Tom, I am about average in that I'm making the usual amount of mistakes and progressing at about where I should be, whatever that is. According to the syllabus, as indicated, I should be a lot ahead of where I am.

A few other realities have come into play during these ten hours which are worthy of consideration. One of them is that the typical instructor seems to be overloaded with students; he makes money in terms of the number of hours of flight time he puts in with his students. As bright and as capable as Tom is, he is obviously unable to remember exactly what it is that I have learned and haven't learned except in a rough sort of way; he makes no notes nor keeps any personal log of what he has taught and what he has not taught. I have since learned that there is a great deal of general dissatisfaction with the economic status of flight instructors; a recent article in *Flying* magazine stated that the average certified flight instructor earns about $150.00 a week, a fact that reenforces the prevailing fact that most flight instructors are just on their way elsewhere in the world of aviation.

Furthermore (and this would apply to most of the country) when the weather becomes marginal (low visibility, overcast) there is that point where the instructor is willing to go up with the hope that he can find some decent weather as against cancelling out the flight entirely and losing the teaching and earning time. By this, however, I do not mean to indicate that Tom or any other good instructor would go up with a student (or themselves) in really bad weather. They won't. If there is anything that these instructors or any good pilot has, it is respect for the dangers inherent in flying and in weather, the medium through which flight takes place. They will all abide by the VFR (Visual Flight Rule) minimums of three miles visibility and 1000 feet ceiling.

One lesson consisted of one of these marginal days where, after we got up, it became obvious that Tom was more concerned than I was about the weather. After we flew awhile and got a report that the weather was getting no better in any particular direction, we turned around and came back. For the first time, I sensed his genuine concern about the dangers of the weather and of the potential of midair collisions. He pointed out that we could not see a plane in front of us far enough ahead to do anything about it and that that is "not good." I'll say! But I still ended up paying for that "lesson" though frankly it was a waste of my money in terms of the anticipated learning process —but perhaps not in terms of an equally vital form of learning.

The major emphasis in the last three or four hours of flight instruction has been devoted to the landing approach system. In most instances what is known as a left-hand approach system is used at an airport. It consists of an established pattern at a certain altitude which all aircraft must fly prior to descent for their landing. What we have been doing is going out to a small, one-runway airport southwest of Chicago, in Frankfort, Illinois. At this airport, Tom demonstrated *once* the idea of entering the pattern at a specific altitude on the downwind leg as it is called, then turning base and then into the final approach for the landing. Just as the plane touches the ground on its main two wheels and the nose gear barely touches the strip, power is applied and a takeoff performed back into the pattern and so round and round you go; it's called a "touch and go." This is no small feat to accomplish. It involves the maintenance of the proper altitude, getting your directions and distance correct and then most importantly, getting that slow flight to the point where your air speed is correct and you use your throttle to drop your altitude, clearing obstacles and making the landing strip. One such day in question, I did it fine the first time around and then progressively worse after two or three attempts. Tom then took over and demonstrated again. I didn't get much better after his demonstration, however.

The next time up we went to the same field at Frankfort, only this time Tom introduced the use of the flaps, used to steepen the angle of descent; the full flap approach will also allow the plane to be landed on very short runways or fields. On the downwind leg, Tom called for 10 degrees of flaps completely forgetting that he had never told me how to operate the flaps nor had we any practice in flying with flaps; we went ahead with immediate flap training. It really didn't come together much that day but Tom promised some training in flap control as part of further lessons.

What seems crucial in all of this is the ability to get that plane down safely. There is no question but that this is a matter of training as well as instinct. You've got to level off the plane about five feet above the runway and then pull back on the wheel in order to, in effect, cause a slight stall, putting the rear wheels down gently (the great dream as yet unaccomplished!) and then letting the nose wheel lower. How do you know when you are five feet off the ground? No instrument will tell you; it is a matter of "feel."

This is not without its direct analogy to flying radio-controlled models. The real threshold is the ability to land the plane. In the case of radio-controlled models, as earlier indicated, this is no easy task since the plane is coming through the air and you have to sense its speed, its altitude, its direction, and where the field is, all at the same time. I am certain that there are individuals, and many of them, who can never land a radio-controlled plane because they can never get that combination of movement and feel right; I'm certain that it is true of full-scale aircraft as well. I think it is generally acknowledged that the critical aspect of flying for the military as well as in something like this private pilot's course comes in the ability to develop that coordination and "feel" for landing. I have it with respect to radio-controlled models. But between the point that I am at now in terms of flying radio-controlled models and my attempts at getting to and over that threshold, there lies in many pieces, many airplanes. It is apparent that you can't do this with a fullscale plane, but the question remains unchanged and, at this point, unanswered; do I have the coordination and the "feel"?

4

Considering the Inducements

It is both curious and interesting to look back over the advertisements that I read prior to undertaking flight training.
Having spent about thirteen hours in the air and probably in
excess of forty to fifty hours of study, even at this stage, some
evaluation having validity can be made. No, I'm not talking
about the kind of advertising gimmickry that shows an individual apparently in flight training with everyone smiling, dressed
in suits, ties, and the like. No one can take that kind of advertising seriously. I have never seen any flight instructor wear a
tie, much less a suit; I think Tom would be rather definitely
embarrassed if he had to wear a jacket with a Cessna emblem
on it.

What I refer to are the basic inducements held out in advertising put out by major aircraft manufacturers in order to
encourage individuals to undertake flight training. How honest
should such advertising be? My answer is that flying is a very
serious business, though learning to fly and the experience of
piloting your own plane can indeed be a thrilling as well as
challenging experience. But there ought to be, I believe, more
honesty than is evident from current advertising materials.

One example of this is the fact that the public will frequently
associate the act of soloing an airplane as equivalent to being a
licensed pilot. Of course, this is very far from the truth. Yet a
number of advertisements have appeared which take advantage
of this misunderstanding by the public and distort the realities
involved. For example, a recent Piper ad reads in part as fol-

lows: "If you can spend two or three hours a week on lessons at an authorized Piper Flight Center, you'll probably solo in less than 30 days." Indeed it is defensible to make that statement on the basis that if you spend approximately ten or twelve hours in flight lessons and if you catch on pretty quickly, depending upon the area of the country and the circumstances involved, you could have soloed. But first of all, the blocks of time are at least two hours for each lesson, about one hour of which is going to be spent actually flying, so you're not talking about three hours a week but more like six hours a week in flight lessons. Furthermore, what about your ground school training? What about all the study that goes along with those flight lessons both at the Pilot Center and at home? Furthermore, implicit in the promise of probably soloing is the play on the misunderstanding that soloing equates licensing and the end of your training.

In an advertising piece put out by the General Aviation Manufacturer's Association (GAMA) called "Twelve Basic Facts About Becoming a Pilot," they attempt to answer what they called the most frequently asked questions about becoming a pilot.

Why should I become a pilot? GAMA stresses the ability to move faster and more efficiently by plane than automobile. I suspect that is not why very many individuals become pilots or at least undertake flight training. It is as individualized and rather unique as each person is who undertakes flight training; I wonder if there are any valid generalizations. For example, there are a number of individuals who undertake flight training in order to actually overcome their fear of flying, their fear of being in any airplane under any circumstances. It might be an extreme way of surmounting your feelings on the subject matter, but apparently there are a number of individuals who do just that.

How difficult is it to become a pilot? Their response is probably more honest than any of the "How I learned to fly and loved it" books that I have read. They state the following:

> As in any instructional process, you have things to learn and skills to master. It is a fascinating experience. But it is not particularly difficult. It can be learned easily by practically anyone who is willing to invest the relatively modest amount of time and effort.

There are two aspects to becoming a pilot. You learn to fly by actually handling the controls of the airplane yourself. Under the supervision of your certified flight instructor, you will learn how to take off, land and fly cross-country. The other aspect of your training takes place on the ground where you cover flight planning, navigation, radio procedures, flight rules and regulations and weather. When you have acquired all this knowledge and practiced the skills involved in controlling an airplane, you will be ready for your private license, which is official recognition that you are prepared to be a safe, competent pilot.

No question about it, there are things to learn and skills to master. It is a fascinating experience but I challenge that statement which declares that it is not particularly difficult. It *is* difficult, and when I say that I think I am honestly reflecting not only my own experience but that of everyone else whom I have spoken to who are at least close enough to the learning process to have remembered how difficult it was. It is probably fair to say that to discuss this matter with a pilot who has had his license for ten or twelve years, you are not going to have an individual who can honestly remember much about their own flight training.

Furthermore, flight training today is much more systematized, regulated and demanding in a number of respects than it was in the "good old days" when a pilot might take up an interested individual and informally teach that individual flying skills with the student reading a book or studying some regulations before taking the ground tests followed by the in-flight testing. At least they stopped saying that it's as easy as driving a car or that it can be learned by anyone who can drive a car. But on the other hand, they do not indicate that some individuals simply cannot be successful at it and that the only way that you'll know whether you've got the talent and are willing to make the very substantial investment in time, is to try it. It simply is not true when they state, "It can be learned easily by practically anyone who is willing to invest a relatively modest amount of time and effort." Here I am, a professional instructor for all of my adult life teaching in a professional school and I can honestly state that, for a serious student, the time involvement in terms of studies and flight training, at least for the time from the initiation of the learning process through obtaining the license, constitutes an investment of time and energy comparable to that of a beginning law student. And ask any beginning law student just how much time and effort they devote to acclimating themselves to law studies and the rigorous demands in terms of time and effort involved. I very seriously contend that the two are analogous. Therefore, to advertise easy learning by practically anyone tied to an investment of relatively modest time and effort is simply untrue. It is as untrue for a beginning law student in that first semester as it is for an individual undertaking pilot training who takes the matter as seriously as the undertaking deserves.

How long does it take to become a pilot? GAMA gives the minimums in terms of the requirements to receive a private pilot's license. Very few individuals, however, ever become proficient enough in their flying skills to qualify for a private pilot's license in those minimums. Tom tells me that he has never had a "minimum" trainee and doubts whether anyone else has ever really qualified and gotten their license within the specific minimums referred to; the exception would be where training is obtained at a residential flight school where it can be pursued on a full-time basis.

Further down under this same question, GAMA states, "Statistics indicate that the average student pilot will complete the requirements between four and six months. However, some finish it sooner and others take longer." That's a relatively honest statement though once again the "half truth" is present in the absence of an honest discussion of the amount of time that the individual will have to devote during that interval in order to finish in that stated period. It is that crucial factor of time, both in flight training and ground training and study time that is the significant variable. As earlier indicated, that time variable is greatly affected by such things as weather conditions and the normal demands of the average individual's work and family life; these must be taken into account in assessing how long it is going to take from start to finish. It can't really be determined how lucky you will be in terms of weather conditions and other factors that might cause cancellation of lessons or postponement of training. If you're talking about the average climactic situation within the continental United States, you're talking about variations where the weather is going to be plenty bad and the chances of hitting it right for a continuous series of flight lessons becomes very slim at certain times of the year.

How much does it cost to become a pilot? The cost of learning to fly is very high. As of this writing, a reasonable figure for the cost of instruction to license would be approximately $2000 to $2500. Once you have soloed, the way you build proficiency and competency is in time in the air. That, in turn, costs a lot more money in plane rentals and fuel.

How about the physical and other qualifications? The minimum age requirement for getting a private pilot's license is seventeen, but you can start when you are sixteen years old. There are no maximum age requirements. While the FAA requires a physical every two years after you receive your license, to my knowledge there is nothing more required of the 70 year old who would continue to fly than the 20 year old renewing his license.

Are the tests difficult? Their response is to explain that two tests are required, one is the written examination and the other a practical flying examination. As to the flying examination they conclude, "... It is not hard to pass the test." The same impression is left with respect to the written examination, in effect stating that you've done it all before and so there's no problem involved. As already indicated, in fact, preparation for the written examination in terms of ground school studies are long and difficult for the average individual because it is covering subjects which simply are not within the gambit of everyday knowledge or of the working life of the average individual. For example, other than watching the local forecaster talk about tomorrow's weather, how much involvement with the science of meteorology does any individual really have? The truth of the matter is that the student pilot has entered a new world, literally a new environment in the learning process involved in both flight training and ground school. It is as much of a new environment as, once again, that freshman law student entering his law studies. Just as all of us have seen and perhaps may have even ridden in a light aircraft or a commercial plane, that fact has not in and of itself really introduced us to the world of flight... this new and quite fascinating but also quite different and difficult-to-understand environment.

In a sense, the promotional blurb is correct; the tests themselves are not supposed to be very difficult. For example, it is a well-known fact that most individuals do pass their flight tests on the first try; likewise, there are books that are readily available and "crash courses" (they last one weekend) that guarantee that you'll pass the written portion of the test. From that standpoint, from the standpoint that the written test can become almost a memorized cinch and that the flight test standards are not that high, indeed the tests are not difficult. But the larger question, the question that presents itself to the person who really wants to fly with proficiency and safety is not in the testing process but in the learning process. The learning process involves much more than the tests would indicate; the learning process involves far more learning in breadth and depth than the testing requires because competency is not really demonstrated by the testing process, at least the kind of competency that one wants to have for his own safety and the safety of others.

Is flying safe? The answer as given by GAMA can be readily imagined; airplanes are safe, people who fly are safety conscious and most pilots are never involved in a "flying mishap," but the latter being an interesting euphemism for potential deaths. GAMA holds to the inducement, "But a well-built and well-maintained airplane in the hands of a competent and prudent pilot makes flying safer than many other forms of transportation." Not only do they qualify the airplane and the pilot but they fail to state what other forms of transportation that well-built plane and competent pilot are safer than. As has already been indicated, anyone who thinks he is engaged in a really safe undertaking when becoming a student pilot is simply kidding himself; you're better off on the golf course by far. Flying is complex and aviation errors are unforgiving errors. Certainly there are plenty of instances of misjudgments resulting in less-than-fatal or even serious personal injury, but collisions in the air, are a very real and dreaded potential.

As an example of the potentials of a midair collision, I turned to a 1977 case involving a law suit brought by the parents of a student pilot against the United States of America (because the U.S. operated the airport control tower) his flight instructor and the pilot of the colliding plane.

It was apparently a beautiful Southern California day about nine o'clock in the morning at the Santa Monica Airport. Traffic in the sky was reported as light to moderate. The Cessna 150 with the student and his flight instructor had requested permission to "work the pattern," that is, to fly the entire pattern in order to practice touch-and-go landings. The Cessna received permission from the tower and was then in the downwind leg of the pattern when another pilot, flying a Piper, entered the pattern without radioing the tower that he was doing so. The court presumed that his radio had failed but pointed out that Federal Air Regulations specifically prescribe how a pilot and a tower can communicate in such an event.

The court went on to recite the facts as follows:

> Prior to the collision, the Cessna 003 was climbing from beneath into the rear of the Piper. Upward vision of the occupants of the Cessna was blocked because of the aircraft's high wing configuration. Blind spots can be compensated for by head movements and aircraft movement.
>
> At the time of the collision both aircraft were heading on a course of 030 heading in the general direction of the sun.
>
> Neither the Piper nor the Cessna 003 executed any evasive maneuvers prior to the moment of collision. They failed to see each other.
>
> At 9:07, the tower called Cessna 003 and asked it to look at a fire on the ground to its left. The controller unwittingly was asking the ill-fated aircraft to examine its own burning ruins.

The court then went on to assess fault in the matter pointing out that with respect to the traffic controller, all he had to do to clearly see the entire pattern was to move a quarter turn of his head. Apparently, he and the other individuals in the control tower simply failed to see the collision course. The court clearly indicated that the controller could have helped avoid what was about to happen if he had looked. The United States government was assessed a percentage of fault in the amount of 20 percent, the pilot who entered the pattern without awareness of the tower (who was also killed) was assessed 45 percent of the fault; 25 percent of the fault was laid upon the student's instructor and the student himself was assessed a percentage of 10 percent of the fault.

It may be well argued that the recitation of such a collision and the deaths of all involved is no different from what could be read in a hundred automobile accident cases. The difference is that the average person knows well the high death count from automobile accidents; the general public also knows that there are millions of accidents each year in which there is no injury or only minor injury to those involved. On the other hand, the aviation industry, while bemoaning the fact that every time a light plane crashes, it is news (and thus distorts the reality of air traffic safety) at the same time go to an unwarranted extreme in failing to realistically apprise the prospective pilot of the safety factors and injury and death potentials involved in flying light aircraft.

A much more honest source for evaluation of flying safety can be found in all magazines published in this country and directed toward the pilot population or those just generally interested in the development of flying and piloting techniques. Surveying any month's offering of these magazines must lead one to the conclusion that there is a deep, abiding, and realistic involvement by the piloting community in the safety aspects of flying. Typical articles will deal with such safety factors as the dangers involved in icing on planes, management of fuel consumption, tips on forced landings, and dealing with the pilot's persistent enemy, the weather.

Almost all of these magazines have a column devoted to the experiences of pilots in meeting and sometimes overcoming potentially disastrous situations. The magazine articles are honest, far more honest than any other readily available source, especially considering the fact that these magazines obtain their income from advertisements by airplane manufacturers and rely upon the readership of pilots. Yet they do, for the most part, "tell it like it is" in terms of the unbelievable range of safety pitfalls involved in flying. It is more than apparent that good piloting (translatable into safe piloting) involves undertaking and successfully completing an adequate course in pilot training followed by a great deal of experience, while that individual builds up the ability to handle the demands of the constantly shifting variables when man enters a flying machine.

It is just too easy to ignore this matter of flying safety. As has been well expressed, the air as a medium of human existence is terribly unforgiving. Human error in the form of pilot error in a light airplane is a deadly combination. It is also fairly evident that not a great deal has been done or is being done to protect the pilot or passengers from the impact of pilot or mechanical error resulting in a crash. All efforts seem to be devoted to improving the quality of the aircraft itself but little, if any, energy seems to be devoted to making flying safer in terms of the crash potential to the pilot and passengers. By that I mean that while airplanes were the first to have required seatbelts, in fact, practically nothing more has taken place in terms of development for the safety of the pilot and his passengers. The requirements for the installation and use of shoulder straps is just now coming into the regulations. There is very little, if other devices to cushion the impact of a crash. Of course, the obvious is too readily apparent; at this stage in our technological development, there is nothing to stop death from a serious crash situation. Therefore, reliance has been on the simple seatbelt, which has been found practically useless in terms of restraining the body upon impact.

What about insurance? is part of that material asked and answered by GAMA. The response given is the following, "If your policies are fairly recent, chances are you are already covered." This is entirely misleading because, initially, there is no disclosure as to what *kind* of insurance is being thought of or discussed. This led me to investigate whether I was insured as a student pilot. As the owner and operator of an automobile, I knew I was not insured by that policy because it is specifically directed to the operation of a motor vehicle by the insured party. I also knew that my homeowner's policy would not cover me because every homeowner's policy specifically excludes liability for the operation of an aircraft. In fact, as this is written, there is a case pending on appeal in which a court excluded from coverage the piloting of a radio-controlled model under homeowner's coverage because the judge ruled that it was within the definition of the "aircraft" exclusion that appears in the homeowner's policy. What I thought would cover me would be an umbrella policy which I carry in the amount of a million dollars, thinking that without an underlying coverage, this coverage would pick up the liability for any loss up to a million dollars. I was wrong. A careful check through my insurance agent and a re-reading of the policy indicates that flying an aircraft is excluded from that umbrella-type insurance.

Now it should be understood that what I am talking about is the potential liability in the event that you fly a plane and crash it into another person's property, another plane in the air, or by virtue of your flying do bodily harm to anyone on the ground or in the air. Plainly put, if I misjudged a landing and smashed into a home or a series of automobiles and assuming my estate would be liable, was I insured against that liability? The answer that became abundantly clear as I reviewed this insurance situation is "no." I then took a careful look at the contract that I sign every time I fly as a student. The contract has a box where, if initialed, (as in the case of renting an automobile) you can get a collision damage waiver so that in the event that you harm the airplane *you are flying in,* you are insured from the first dollar of loss. If you fail to initial that box, there is a potential $1,000 liability.

As a student pilot, the outfit with whom I was taking lessons picked up the cost of that collision waiver. Great! That means I'm fully insured with respect to any damages done to *their airplane* but more importantly, what about the damages you can do with the plane! There is no coverage for that aspect of potential liability offered when you undertake lessons or otherwise rent a plane. What must be understood is just how great the potential is for wiping out your estate and in effect leaving your family pennyless if, as a student or private pilot, you do substantial harm or injury to people or property with that plane. Yet the one paragraph put out by GAMA in response to the insurance problem, ignores this entire matter.

What I did do (and what I would certainly recommend to anyone undertaking lessons in flying do) was to get insurance in the form of an aircraft nonownership personal liability policy. I immediately took out this insurance which, at a quite reasonable cost, gives me the maximum of a half million dollars in bodily injury and property damage coverage. I wish I could have obtained more; this was the maximum available. Thereafter, should I become a pilot and continue flying, a policy is available with higher limits, with the rates going down as the number of hours logged goes up.

Turning to the matter of life insurance, the rest of the information is misleading, if not totally incorrect. The information given leads one to believe that there is no problem in getting life insurance or any other kind of insurance (such as accident insurance which insures you against your own accidental injury or death) as a private pilot; they seem to indicate that it no longer makes any difference.

In terms of accidental injury-death insurance, it is now standard procedure in the insurance industry to collect a substantial premium where the insured is a private pilot. For example, many if not most companies will charge up to 200 percent or more additional premium when you are a student pilot or licensed pilot under the age of 27. Beyond that age, insurance premiums involved an additional premium over a non-pilot premium based on the number of hours of flying. For my category, that is, a student pilot over the age of twenty-seven, the additional premium percent is listed as "one hundred fifty percent up." This higher premium quoted also carried the following language concerning those higher premiums, "The above extra premiums cover the usual hazards of flying in the U.S.A. When the accident record, the type and amount of flying or terrain covered suggests an unusual hazard, refer to your operating office." This language is directed to insurance agents and brokers, but speaks pointedly to how the insurance industry feels about insuring pilots for accidental injury and death insurance.

The typical application for accident insurance or life insurance will routinely question, "Has any person to be insured flown as a pilot or crew member in the past five years or contemplated doing so in the future?" If the person taking out the policy checks "yes," then you can figure on the probability of a quite substantial premium for that insurance coverage. The surcharge made will be, as to life insurance, on a premium of a certain amount over the normal per thousand rate. In turn, the surcharge reflects the amount of flying experience; again, the higher surcharge being related to the lower experience. Since the question covers "contemplated" flying, the best that can be hoped for is that the person undertaking flying has sufficient life insurance or can obtain additional life insurance from those few instances where a company does not demand a premium for the private piloting of aircraft.

To conclude this section, I would draw attention to a statement that Cessna puts on their postage meter stamp which states, "If I can fly, you can fly." For some time, Cessna used to advertise that, "If you can drive, you can fly." They backed away from that one in recognition of the fact that skills in driving do not necessarily correlate with skills in flying. But have they really moved away from the broadest form of inducement when they state that, "If I can fly, you can fly?" This whole idea belies how complex the whole business of flying is and how much safe flying depends upon skill, practice, good judgment, and luck. The absence of at least good judgment, practice, and luck was evident in a recent report of a fatal crash retold in the October, 1977, issue of *Flying*. The article was about a relatively new pilot and was based upon an actual National Transportation Safety Board Accident Report. The article reviewed the communication between the pilot and various flight communication offices along the intended route. The individual involved got into some very bad weather and got disoriented. Despite all the efforts of those on the ground, he lost control of the plane. His last recorded transmission was, "I'm descending, descending, Oh God, (unintelligible) need help, Oh, somebody help me. I'm gone. I'm gone." There was nothing wrong with the airplane; it was simply a case of an individual who should not have been flying where and when he was, who made a poor judgmental decision by staying aloft in adverse weather conditions for which he was neither sufficiently trained nor experienced.

Interestingly, in the same issue of the magazine, the Associate Publisher wrote an article entitled, "What's Wrong With Flight Training?" His basic answer was that there is a lot that was wrong and that needed corrective action. The statistics he quoted were not quite what I heard at the Cessna Pilot Center or from Tom. The author states that six out of ten student pilots lose heart and drop out before obtaining a private license. He also indicates that as many as *80 percent* of those interested in aviation to the point of taking lessons and obtaining a medical examination and student license eventually become sufficiently disenchanted to leave flying before becoming an active member of the private pilot group. He asks why and answers basically in terms of the poor quality of teaching. He asserted that the average wage of a flight instructor is $150 a week and is looked upon as a mere stepping stone to doing something else in aviation, a way of making a few bucks while you're on your way somewhere else. To my astonishment, because I felt that it was exaggerated, when I cited some of these statements to Tom, his response was that of entire agreement; he was just pacing time, though it was obvious that he had sufficient natural teaching talent to impress me, a professional teacher, with his teaching ability. He confirmed that the $150 a week figure was about right over the spread of a year, given weather conditions and the amount, apparently, of his share of the cost to the student.

So we are told, "If I can fly, you can fly," but what is not told belies the simplicity of this statement, a statement in and of itself simply not defensible.

5

Toward Soloing

With about twelve hours of flight instruction logged, the last of these hours proved to be the best in terms of my performance. It was obvious that Tom was pleased with the way I was handling my touch-and-go landings at Gary Airport. I was able to get down to the appropriate slow airspeed, hold that speed, line up nicely with the runway, and get the plane into a nice glide angle down to the runway every time. I was still experiencing difficulty in judging just when to bring the nose up from level flight into the "flareout" position. I was either a little too high or I hit the landing strip "pancake" style, instead of the light touchdown on the main gear with the nose held off the ground until the proper instant. This important flight task is difficult to articulate in writing but it is even harder to actually perform. You have to set up the proper glide angle as you approach the landing strip, then level off about five or six feet above the landing strip past the numerals (they tell you what landing strip you are on). Then, as the plane sinks because the air speed is down and lowering, you have to time that pulling back of the wheel so the touchdown, as indicated, is on the main gear, then gently lower the nose. It was practiced over and over and, while my takeoffs from the "touch-and-go" and getting into the pattern position were fine, that last fifteen or twenty seconds as you fly "down the runway" call for the greatest amount of perception and coordination. They were, however, landings which, while not "respectable" to an accomplished pilot, certainly were within the range of a safe landing.

The truth of the matter is the day was just "perfect." The wind was practically calm and the visibility excellent. How much good performance depended on those conditions remained an unanswered question. Furthermore, the landing strip at Gary was long, wide, and quite visible. But again, Tom was pleased and that certainly was encouraging.

Something rather unusual, and quite amusing occurred as we approached Gary and I called to inform the tower that we were coming in for "touch and go's." The response of the tower was to the effect that the tower controller would have liked us to "get lost" and told us so; he announced that he was "sick and tired of pilots coming over for touch and go's." This rather astounded Tom who got into some conversation with the controller, who finally relented. Later Tom explained to me that this was not a Federal Aviation Administration tower but rather one owned by the City of Gary and therefore the standards are not what one would expect at an FAA tower. I said I thought it was their job to do just what they were supposed to do: run a tower where pilots would "do their thing"—with the right to do so, so long as it was within the appropriate flight rules; Tom agreed, but indicated that in this unusual instance it was a matter of personality as well as the absence of a federally regulated tower.

Two events on that same flight rather frightened me. The first was when, without warning, Tom jammed the throttle to the fire wall (giving the plane full power) and pulled up on the wheel sending the plane practically straight up. Without a word, he had reacted to seeing a plane which could have been on a dangerously direct heading for midair collision. The more I read about this subject, the more I have found that next to pilots flying in weather that they have no business being in, the midair collision is the dreaded hazard of flying. An interesting comparison holds in flying radio-controlled models. There, a midair collision is extremely rare even with what seems to be a sky full of models. It was only after three years of flying radio-controlled models that I saw my first midair collision, and it happened that one of the planes was mine! The incredible part of a midair collision in flying radio-controlled models is that, with only the very rarest of exceptions, there is nothing that can be done to avoid a midair. It's just too late for anyone to react. You do get training in how to avoid midair collisions with full-scale aircraft, but basically it's a matter of keeping your "head on a swivel." Tom reacted quickly and had taken us out an an apparent danger. I questioned him about his maneuver since according to what I read, the called for procedure was to pull off to the right; Tom's response was that he could not be certain the other plane wasn't in a flight path that, pulling off to the right, would not have brought us onto a collision course.

The other "frightening" aspect of the day's work was something quite different. Tom asked whether I had had the medical examination for my Student Pilot License. I hadn't as yet, though I had a sheet of FAA approved doctors and had made a few calls trying to get an appointment. Tom indicated that it was time that I did get that examination so that he could endorse the Student Pilot License which is on the back of the medical form in order to authorize me to solo the plane within a certain designated area. The "frightening" part of this is that apparently the day's performance in flying had reminded him that I was near the point of soloing. The whole idea of being that close to solo flight was in itself a frightening thought. As it happened, I was months away from my first solo.

Tom's remark set my mind to imagining what it would be like, when it would take place and just what could happen. I kind of imagined that it would be on a flight down to that tiny Frankfort airport; that little uncontrolled strip of paving where, in a few hours, he would simply step out of the plane and tell me to take off, go through a pattern, and bring it back down again. Only time would tell whether my imagined idea as to when and under what circumstances that solo flight will take place would prove correct.

My ground studies continued. The Cessna course is divided into eleven large lessons for their Private Pilot Course, each lesson subdivided into three sections. It is no easy task to keep up with the studies. After chapters 3 and 4 came my second quiz. This quiz, like the first, consisted of twenty-five multiple-choice questions. No particular time limit is set though I think the average student would take twenty minutes to an hour in taking either of those exams. On the first test I got six wrong out of twenty-five which the young lady who graded the test said was "quite good;" it didn't sound so good to me at all. Tom reviewed where I had gone wrong in my responses and one basic problem was obviously reflected in at least a couple of the incorrect answers; my lack of "direction sense" plagued me. I've never been one to know whether I was going north, south, east, or west.

A compass had always been a mystery to me and it was evident that I was going to have to overcome a real problem here. Even on that good flight day, flying back to Midway presented that challenge of figuring out the pattern configuration for the proper landing, which in turn involves "sensing" direction and doing so in quick order. It gets pretty involved to describe, but what it amounts to is that a person with a very good sense of direction has a much easier time in this important area of flight instruction. I understood that this could be a "make and break" in terms of my own success as a student pilot.

Returning to the matter of ground studies, I estimate that in preparation for that second quiz I must have spent about twenty hours of study time. It did pay off in the sense that I achieved a score of twenty-two out of twenty-five correct, but it took that kind of study and restudy in order to gain enough knowledge to do that well. The questions are not in the least easy and are obviously of the level demanded by the FAA (or higher, as I was to learn) in the written part of the private pilot's license examination. This last week the studies have gone even slower moving through chapter 5.

Whatever cliches you want to use about weather, there is nothing easy in learning about the world of weather as an academic study. Once again you are confronted with so much that is totally new. Once again, there is a meaningful analogy in this to the study of law. We live in a society in which the law as a human institution surrounds us, envelopes us as does the air we breath. In the same way, the weather is part of our daily lives. Yet, as to both, unless you are involved in the study of either of these environments, the weather or the "legal" environment, they are pretty much taken for granted, perhaps equally complained about. The entire study of weather phenomena and the reporting system involving a complex shorthand system of symbols and letters was absolutely overwhelming; it's all so new and quite complex. I took my tape recorder and recorded a number of the tapes used for these weather lessons and listened to them over and over as I drove in my car simply to become more familiar with the language and with some of the ideas involved. It became obvious that it was going to take more, much more in the way of just good, hard study to reach that level of proficiency necessary to handle that next Cessna examination and ultimately, if it came to pass, the private pilot's license examination with its weather questions. How can you escape it; the airplane moves through the air and the air is "weather." Adverse weather is the single greatest factor involved in fatal flying accidents; this makes the study of weather and weather-related phenomena compellingly meaningful as well as academically demanding.

Perhaps I felt myself more than ordinarily sensitive to the dangers involved in flying. Here again an analogy is obvious; to the medical student, it is well-known lore that the student frequently feels as though he and his loved ones have every (or at least many) of the diseases studied. Furthermore, the study of medicine leads one to feel just how susceptible the body is to so many diseases that the average individual would never have heard of or considered; so many examples of the breakdown of the human body are given, often with such ugly and fatal results. In the same way, as a law professor, across my desk comes a constant flow of reported cases from our courts. Since beginning the flying experience, I have been collecting cases involving airplane crashes in what is called general aviation, as distinguished from commercial aviation. Prior to undertaking flight training, these cases were simply passed over.

Here a case sits in front of me involving a student pilot who was severely injured and sued the United States of America, charging that the air traffic controller failed to warn him sufficiently and properly of what is known as "wake turbulence" from a jet landing in front of him. How can you help but ask yourself how far you are from this situation as you read the court reciting, in part, "At the time of the accident in 1973, plaintiff was a nineteen-year-old student pilot with thirty-two hours of flight time. The accident occurred while he was on his first solo cross country flight from his home field at Winona to Rochester. He was practicing touch-and-go landings at the Rochester Airport."

The plaintiff (student pilot) lost this case because he had training in how to handle wake turbulence and he failed to do so. Thus the court denied him recovery though in part the court added, "Finally, plaintiff contends that the controllers were aware, or should have been aware of his precariously low altitude and negligently failed to warn him. The court admits that it is troubled by this aspect of the case. Perhaps it would have been wise, given the fact that the plaintiff was a student pilot, for the controllers to adopt a 'mother hen' attitude toward him. However, they were not required to do so. . . even though a student, (he) was pilot in command and therefore was primarily responsible for the safe operation of his aircraft." It just happened that part of the chapter I had been working on the week I read this case dealt with wake turbulence and how to avoid same. Needless to say, I went back and read that section over carefully a few times.

Perhaps it is a distorted view of the world of private pilot flying to collect and read these cases, as distorted as that medical student dealing with constantly diseased and injured bodies. But it's just as distorted, I would argue, to take the view that there are no significant risks in flying. There *are* risks and one method by which anyone can begin to feel the "safety factor" involved in learning to fly and in piloting a plane is to study magazines directed to the private pilot. You won't find an issue without "safety tips," stories of near and actual accidents and, generally, the clear realization that safety is central to this whole flying business.

Almost one month expired between flying sessions in mid-October to mid-November, a month where little progress had been made. In fact, "progress" is inaccurate; I had gotten worse in my landings since that excellent session at Gary airport and I was becoming somewhat discouraged. No doubt about it, I was now well behind the "average" student; I should have soloed by this point.

To make matters worse, a combination of weather and instructor-cancelled sessions, mainly adverse weather conditions, explained the delay. It was disheartening. It seemed that every day that had been picked for a flying session turned out to be the windiest, rainiest, or foggiest day. No flying can take place below the VFR (Visual Flight Rules) minimums and those minimums were simply not there.

The truth of the matter was that in each case, only an instrument rated pilot would have flown in those weather conditions. This is another reminder of the very real limitations that exist with respect to the average private pilot; only 35 percent of all private pilots are instrument-rated pilots which allows them to fly under what are known as IFR (Instrument Flight Rules) conditions. Becoming instrument rated is the next big step after you have your Private Pilot's License and have put in a great deal more in the way of flying time and studies, eventually passing a written as well as a flying test in order to get that rating.

I continued my ground school studies, delving further and deeper into aspects of flight phenomena. When I took another test covering a portion of ground school instruction, I was tripped up by some literally tricky questions where I felt the obvious intent was to catch you rather than to test your knowledge. Tom explained that they put these questions into the test in order to alert you to the fact that the actual FAA test has some questions that are precisely that way—they are trick questions, though I must add that the FAA denies that they have any such "trick" questions on their written exam. This was especially meaningful to me because as a teacher, the kind of examination that would contain "trick" questions goes against the grain of everything I believe is correct in terms of testing a student's knowledge of any subject matter.

It was past mid-November when the weather and a scheduled flying session came together for me. Tom felt that this lapse of time might indeed be a very good thing; he reasoned that sometimes it's just good to get away from flying for a while (for me, evidently from some bad habits in flying) and then take a new look at the matter. He had made it abundantly clear that until I could get those landings down, there was no use or need to get any practice doing anything else.

I found myself both comfortable with the idea of returning to flying and quite anxious to see whether that lapse of time would make a difference. We got off the ground and out of Midway and its traffic quite smoothly. My control of the plane during the taxi was very good. To top that, I found the destination airport with no difficulty and descended into the pattern at the correct pattern altitude, again without a problem and in fact quite efficiently. Then for some two hours I went through the touch-and-go exercise to develop the ability to land. This time, the exercises went incredibly well. For the first time I was able to get on that runway and touch down with a minimal "thump," indicating I had put down the main wheels on the ground gently before lowering the nose. For some reason, everything seemed to come together. The landings were consistently good and for the first time I began to feel physically and mentally tired after going through the pattern more than fifteen times.

Then Tom said, "Let's make a full stop out of this next one." After the stop, he turned to me and said, "I think you're ready. Give me two touch-and-go's and then come to a full stop and pick me up here. You've got your medical with you haven't you?" Before you can solo, the instructor has to endorse the student as ready for solo flight on spaces provided on the student pilot's license. It had been resting patiently in my wallet for months. His last words to me were that I would feel something of a difference because his weight would not be in the plane, but otherwise I could do just what I had done before. He asked if I was too tired to try it, and I responded that I was not. In a flash he was out the door waving to me as the door slammed behind him.

With no hesitation I taxied the plane over to the end of the runway, scanning the skies, I got on the radio to announce that I was taking off from Frankfort on Runway 9. Frankfort is a small, uncontrolled airstrip consisting of just one runway. Naturally, my first feeling was, "Well, this is it; you're on your own." In the rush of events, I had forgotten that I had "predicted" my solo flight would be from Frankfort.

The amazing part of this experience was that I knew I would not be nervous or concerned, and I wasn't. I am certain my heart beat did not increase; I was totally unconcerned that the moment of solo flight had come. My only thought was the need to do it right, to get the landings correct and fulfill the trust Tom was exhibiting in allowing me to solo.

So up I went and indeed immediately found the lack of

Tom's weight really did make a difference in the plane's performance; it got off the ground faster and was more maneuverable. It must be kept in mind that these trainer-type single engine planes weigh about 1,600 pounds empty, so 165 pounds less makes quite a difference. The two touch-and-go's were, according to my ratings, "fair" and I thought the best was the third landing where I had to come to a complete stop, then turn around and pick up Tom. He surprised me somewhat saying he thought my second touch-and-go was my best of the three landing attempts but I had done a decent job on all three. I was delighted!

We headed back toward Midway, taking up the exact opposite heading that brought us to Frankfort. At the appropriate time I called Midway Tower and advised them I was landing. In response I was told to report, "a right base for runway Four Right." I asked for a repeat on that command and got it. I had to scramble through my thoughts to try to visualize what I had to do. This was an unusual approach because usually a left-hand pattern is called for and the controller was calling for a right-hand approach. It also meant he wanted me to report to the Control Tower when I was on the base leg of that approach into Midway. Tom gave me some advice on where I ought to be at what time on that base leg but I also sensed what was intended. I got on to that base leg and reported to the Tower and was cleared for landing. Tom said I was a little bit low and so I added power to get higher. The landing itself was decent enough and I turned left on a taxi way, crossing Four Left. I reported to Midway Ground Control telling them where I was and where I was going and parked the plane. Larger airports such as Midway have two runways for each direction, a Right and Left, and with the quick speaking voice used by controllers, it is easy to miss which one they mean.

Needless to say, I was euphoric. The solo had been accomplished and that was great. I wish I had as much comfort in that soloing at Frankfort as I must have in terms of handling the demands involved in coming into Midway. The place is so frightening, so demanding in terms of realizing where you are and what you have to do to obey the air traffic controller's directions. This is something that has to be worked on until levels of competency and comfort are achieved.

Back at the Aviation Service office, Tom congratulated me and there were smiles all around as the office personnel appreciated that I finally finished my solo flight.

I asked myself just what had made the difference? Was it simply the passage of time without flying? Perhaps. But there was another factor. At a law school faculty lunch, I was sitting with a faculty member who I knew was an experienced pilot. I told him about my difficulty in getting the landing technique down and he asked me what I used as a visual reference as I descended on the glide path close to the runway. I indicated that I had looked straight ahead down the runway—"Ah hah —that's where you're wrong," he responded. "Glance out the *side* and note the distance from the ground; try to develop your judgment in that way so you know how far your wheels are off the ground before you level off and begin your flare (pulling back on the wheel so as to raise the nose and get the main wheels on the ground first).

I thought about the suggestion and tried it at Frankfort that solo-day. That recommendation plus Tom's advice to line myself up (and not the plane) with the center of the runway seemed to work. Once you "have it" in terms of landing, it seems to work every time in the sense that you won't be too far off on any landing attempt. Some people say that once you have perfected the technique, it never leaves you. But I think that is an oversimplification of the process. After all, you have to be at the right altitude, keep the right airspeed, be in the right position with the plane on every "leg" of the landing pattern, be alert for other aircraft, obey the controller (if there is one), use the flaps correctly, and touch down correctly in the right place. Considering the correct "balance" of factors and techniques involved in a landing, even perfecting the technique can be very complex and demanding.

The next day I looked at my pilot's log where every hour of flight time must be recorded. You may read that the "average" individual solos after some thirteen hours of dual instruction—I had about thirty-four hours before my solo. The long hours did not diminish the feeling of true satisfaction and accomplishment that went along with the entry Tom made of that first supervised solo. No question about it, it was one of the unforgettable moments in my life.

6

From Solo Through
Written Examination

There are those who will argue that it is not wise to stick with one instructor exclusively, but rather to "gypsy," that is, to go from one instructor to another during flight training. The advantages they propose include not being locked into one person's particular flying habits or techniques as well as the challenge of meeting the demands and expectations of a number of instructors. Most students, however, prefer working with one instructor for a number of reasons. Included are the need for continuity of training (he knows where I am) and the benefits of a personal relationship (he knows me, my foibles, my needs). Learning to fly is a deeply personal experience and most of us need the comfort of working essentially with one person in whom we have confidence and with whom we build our flying skills.

Certainly it is the height of folly to stick it out with an instructor that doesn't "groove" with your personality or taste. The human "chemistry" has to be right between student and instructor; the subject matter is too vital to life, limb, and pleasure to allow personality clashes to interfere with the learning process. The cockpit space is small and physical proximity necessarily close; that's another reason why the "mix" must be right. Do you know how it will be when you first meet or are initially assigned an instructor? You won't; once it becomes evident (as it will very quickly) that it isn't working, cut it off, get another instructor and go to work again. Unlike other learning situations, what is at stake in learning to fly is just too important to allow hurt feelings to stand in the way of change.

Having had all flight training with Tom, it was therefore with some trepidation that I scheduled myself to fly with Kerry Johnson on Tom's day off; I was anxious to keep up the momentum of success since the Frankfort solo. Kerry seemed a little older than Tom, but not much more; he was, however, much more experienced as a flight instructor having taught flying for some years at Southern Illinois University. When he told me about this, I thought to myself how great it would have been to have had flight training as part of my university education. In truth, however, it is not just a matter of another "course;" you have to be enrolled as a major in aeronautical studies. Interestingly, the statistics of the main campus of the University of Illinois indicate that, among freshmen, the highest flunk-out rate, by far, is in aeronautical studies.

Kerry knew about my recent first solo and my slower than usual flight progress. We flew on the day after a rather heavy snow had fallen on Chicago and I wondered whether the runways would be cleared. They were and Kerry directed me off to Gary for touch-and-go's, if we found their active runway clear. I flew over to Gary in one of the few Cessna 152s they could get running in the cold weather, number 24337. I did most of my flight training in this yellow and white bird and got to love the particular feel of this plane.

I got into the pattern at Gary and did a couple of laps around with pretty good results, enough so for Kerry to suggest that I make a full stop on the next time around and let him off so I could solo this pattern. So a full stop it was, out he hopped with a "good luck" and I taxied out to the runway for take-off. This soloing was different to the extent that I would be working with a tower as I went around the pattern. My first time around went well. On the second, I noticed (and heard) another plane getting clearance to take off on a runway intersecting the one I was landing on; the wind was very light so that this procedure could be used. My instructions were to report on the base leg of my pattern to the tower for clearance to perform the touch-and-go. This I did and immediately heard the tower abort the takeoff of the other plane; meanwhile, I completed my touch-and-go. The tower then explained to the other pilot that my plane was piloted by a student pilot. I later asked Kerry, who had gone into the tower during my solo, how this had taken place. He explained that the controller should have realized what he was setting up; I was taking assurance that I had not been the cause of the problem. Kerry also told me that the controller had asked him whether I could understand if he instructed me to do a "360" on base; Kerry had to answer that he didn't know because he wasn't my regular instructor, so the controller had aborted the take off of the other plane. I had to ask myself whether I could have obeyed that command to circle on base, since I hadn't done anything like that before. It's a common procedure, but to that point, I simply hadn't been asked to perform it during a pattern.

Working with Kerry had been an excellent experience from many angles. It was a notch in my confidence ladder to have done decently well with him. He also had taught me some new techniques, not the least of which was to suggest I start moving to full flap landings; he later apparently discussed this with Tom who reported that Kerry thought that I did, "Okay, but a little flat on the touch-and-go's," and Tom went on to discuss the need for training utilizing flaps. Now my logbook read, "second supervised solo."

The flight training appeared to be progressing well, but there was another plateau facing me that literally gave me sleeplessness over a period of weeks. Flying out of Midway was no problem; flying to some neighboring airport south or west of Chicago could be accomplished. Doing the touch-and-go exercises were also manageable, but getting back *into* Midway terrified me. The emphasis on "into" is intended to convey the fact that finding the airport was less of a problem than handling the demands of the controller as he barked orders with incredible speed. And then there was the problem of directions. Midway has its runways in a rectangular configuration which makes direction sensing more difficult. Over and over I "flew" into Midway—in bed, at the office, as I drove. At times it seemed to infringe upon my classroom teaching as well, so involved was I in this matter of handling getting back into the traffic pattern at Midway and safely onto the ground. Despite all the times I had done this with Tom, I could not

achieve any measure of comfort; it seemed I was always misunderstanding the controller or ending up on the wrong corner
of the field or setting up for a landing on 22 Left instead of 22
Right. This was anything but funny—it could be just plain
dangerous.

To compound the matter, Tom indicated it was time for me
to fly solo from Midway and return to that home base after
practicing at some outlying airport. I can't remember how often
I thought, this is the time to quit. Why push? You're scared
and lack the confidence, perhaps skills, to handle the matter. I
had visions of everything going wrong. Midway was where anything could go wrong because even on "off" days and times,
there always seemed to be so much traffic. Furthermore, this
huge airport is entirely surrounded by commercial and residential property right up to the airport boundaries.

Tom sensed, I'm certain, my deep concern. At the same time,
he indicated he was confident I could handle the demands. He
proposed that on the next time out, we would fly together to
some outlying airport and then back to Midway where I would
drop him off and then I would go out again to that outlying airport, practice touch-and-go's and return to Midway.

Once again the weather intervened to delay this next step.
Meanwhile, I studied the aeronautical chart over and over,
mentally flying back to Midway.

The day finally arrived, a beautifully clear, cold winter day.
Off we went in 24337 down to Gary and there I did a couple of
touch-and-go's which, frankly, weren't too great. My mind was
fixated on the matter of getting back to Midway. Tom directed
me to fly back, adding that he wanted me to do exactly as we
planned, dropping him off and flying back to Gary for more
practice and return.

At the proper reporting point, I called the Midway tower and
got an expected direction, one that I had mentally practiced; I
was told to report on a downwind leg at the southeast corner for
Runway 22. I did just that and was cleared to land when I
reported on that downwind leg. After landing, I taxied over to
the west ramp where the plane is normally parked, dropped
Tom off, who added as he left that he would try to stick around
the office until I returned so I could tell him how I made out. It
immediately occurred to me that if I could do just that, see him

back at the office having flown the plane and returned, the rest would be pretty obvious—I had made it.

He buckled the seat belt on the now vacant seat and off I taxied back to the active runway, did a short run-up and reported ready for take-off.

Once again I felt inexplicably calm, given the weeks of anticipation. The take-off was smooth and I realized that I was in the air, by myself and heading *away* from Midway.

The trip to Gary was brief and direct; it couldn't have been otherwise given the fact that I had just been there and the day was exceptionally clear; I could almost feel my way right to the airport without glancing at the chart at all. I did three touch-and-go's at Gary, reported to the controller that I was leaving the pattern after the third touch-and-go. Having done so, I was on my way—and on my own back to Midway. Once again, over the proper reporting station I called Midway tower requesting instructions for landing. I turned up the volume of the radio and concentrated with all my might so that I might distinctly understand exactly what directions would be given, hoping that it would be a repeat of the prior directions since I was at the same point and going back to the same active runway. I was relieved to hear the same basic instruction. But then, the tower called me and asked whether I could see an aircraft off to my left rear; I swung my head around and scanned the skies but could see nothing. I reported that to the controller who came back with, "Look again!" I did, and I saw the plane this time to my left rear. I reported back to the tower that I had the "traffic in sight." The tower then stated that I would be number two to land on Runway 31 Left after the other plane which would land on 31 Right.

What a time to change landing instructions! And now where was that other plane and why was he telling me to land to the left which meant I would be crossing over the path of the other plane. Once more, I could no longer see the other plane.

Trusting (and praying) that the other plane had passed beneath and to my right I prepared for the landing on 31 Left. This meant a straight-in approach with no downwind or base legs, just a long final approach which in some respects is more difficult to do properly since you don't have the time to judge all of the factors that must be taken into account that can be more easily managed with the usual downwind, base, and final approach. But I began, pulling out the carburator heat, slowing the engine to 1500 RPM, waiting for the plane to slow down, dropping in 10 degrees of flaps, then waiting for further slowing down and approach to the runway, dropping in 20 degrees of flaps, then over the fence and onto the landing strip. I made it onto the runway with a pretty decent landing flare, but as the plane continued to roll, it began to move off to the left on the slippery runway. I just managed to keep it from going into a mound of snow.

The tower then called and asked where I wanted to park; I responded and he directed me to make a turn and watch out for some emergency snow-cleaning equipment. I made what I thought was the right turn off, only to hear the tower bark back at me that I had made the wrong turn, that it was the next one; I complied, this time getting on the right ramp leading to the west ramp of the airport. As I taxied into the parking area, there was Tom waiting to go up with another student, but delaying this so he could talk to me upon my arrival. The obvious expression on his face and his words of congratulations contained that "I knew you could do it" tone, which I really appreciated. For some reason, despite the bitter cold, I felt nothing but warm all over as Tom helped me push the plane back into its proper parking spot. What a great feeling to have accomplished this flying!

In front of me had sat over 100 law students writing their final examination in my Property Law course. Their concentration was intense; the anxiety level high. The room, though absolutely silent save the quiet occasional movement of an exam booklet page, had fairly bristled with excitement. Faces were strained, utterly involved in the singular purpose of answering the problem presented.

Having gone over the questions and jotting my notes as my "answer key" to be used as I graded the essays, it flipped through my mind—how obvious, how straightforward these questions are; perhaps I had made some of them too easy. How could they (apparently) sweat so hard?

How absurd I was; of course they were sweating and of course the problems were challenging, hopefully not too much so. Had I so quickly forgotten how demanding law studies and examination were? And what about yesterday? Yes, yesterday the weather had finally broken and it looked like I would be flying after having had the weather cancel three lesson appointments.

Out I had driven to Midway, trying to guess what runway would be active (the wind was shifting) and how I would handle the possible alternatives. When I got there, frustration—for my students, the forgotten point of law, the uncertainties as to response. For me, no 152s in operation; therefore, fly a 172, a larger four-place Cessna. I had flown a 172 only once before. But this 172 didn't seem to want to start; the weather, though clear, was bitterly cold. When the battery seemed about to give up, I heard Tom swear for the first time.

We got a "pre-heat" from a friend's homemade contraption —a couple of flexible pipes running from his car's exhaust into the engine compartment of 172 through the cowling. Tom's friend went back to start his plane while we waited for the warm up to take place. Suddenly Tom saw a fire under his friend's plane and we dashed toward it; the friend had jumped out and was pushing his plane for all he was worth to save it from the fire that seemed to be coming from a pool below the plane. The fire engines arrived, but the fire had been stamped out. Back we went to our 172, disconnected the pre-heat contraption and tried the engine. Like magic, it worked, starting right up.

Just as my students had faced the challenge of the examination, so I found the analogous challenge in the cockpit of the 172. Everything was in the "wrong" place; all the dials were in a different position from that of the 152. I kept telling myself, quite in vain, to stop looking in the wrong place for the particular instrument.

We flew out of Midway without any problem, but in the air the plane seemed to behave so differently; it appeared continually to want to climb and I had difficulty finding the right setting for the correct flight attitude. Once again, it was over to Gary for touch-and-go's, but the pattern (which should have been quite familiar) had some new twists with this different plane. There was an element of discouragement here because I realized that the Cessna 172 was probably the plane that I would most often fly once I had my license. It is the most popular four-place airplane around, and more of them are built every year than any other single-engine, four-seat plane.

So this is the plane that I would want to rent or eventually own? At that point the idea didn't look very encouraging, and that 152 looked better all the time. My mind started to rationalize; you don't need a four-place airplane because most of the time people don't want to go with you anyway and isn't it just as much fun in a two-place plane, etcetera.

And so it went, a discouraging flying session; I was trying (according to Tom and he was right) to do something required before completing what had to be done before, thus rushing the needed progressive steps. No different, I suppose you can say from any learning process, whether it be golf, racquetball, tennis, or law.

Perhaps that was precisely the point: the serious learning process sets demands which must be met. They—my students —were struggling with learning the law; I with learning to fly. I had to think of how they would enjoy seeing me sweat this learning process as I know they sweated theirs. On one level, learning to fly was making me a better law professor because I am convinced that if you aren't engaged in some learning situation that demands all you have, it is too easy to forget what serious learning of any kind really involves.

I had determined at this point that I ought to culminate my ground school studies and take the Private Pilot's Written Examination. Since July, when I first began flight training I had been reading, studying the text as part of the Cessna course and doing the problems in the workbook. I had taken all of the quizzes and I felt that with winter at hand, it just made good sense to culminate those studies by taking that examination. You have to pass that written examination before you are qualified to take the flight test as administered by the FAA, so I felt that if the flying was not going too great and the weather was against me, let's get this written text out of the way.

Over the years, the FAA had varied its testing procedures, both in terms of the written and flight test. As of this writing, the written test consists of sixty multiple-choice questions with a passing score of 70 percent to be achieved. The written examination tests a wide spectrum of subject including basic aerodynamics, meterology and weather phenomena, navigation, the airplane engine, medical aspects of flying, and a host of other areas. In theory, the examination is supposed to challenge your ability to plan a flight properly and complete that flight successfully on paper. Actually, however, the problems involved and the questions asked do not really accomplish this goal. What you really have is simply a selection of questions covering the subjects with no real unity involved—just a random sampling of questions from different areas.

When approaching the written test, in a very real sense you are coming to a mountain which you have to carefully ascend and descend, but it's a mountain with an unusual twist—you can also pole-vault right over it. By this I mean that there exists the realistic alternatives either of taking the entire matter very seriously and studying the various areas with depth aimed at achieving and understanding far beyond that which is tested or of simply hurdling right over it in quick fashion.

For reasons where the justification is not clear to me, the FAA for the last few years has published a booklet containing all of the questions (without the correct answers indicated) from which the test questions are extrapolated. That means that you can buy a book from the government for a couple of dollars which contains the six hundred questions that constitutes the universe from which every test is taken; your test is going to consist of 10 percent of the universe of the questions. Furthermore, for just a few dollars, you can get the questions with an answer sheet put out by private publishers or go even further and get a separate booklet that not only answers every one of the six hundred questions but also gives you an explanation of them. To exemplify this situation, I quote from a recent advertisement in a popular aviation magazine:

PASS YOUR FAA WRITTEN
FAA WRITTEN TEST ANSWERS

The FAA is now beginning to publish written test books of 600 questions. Aviation Test Prep publishes answer books for each test book the FAA publishes. Currently, the test books available are: Private Pilot (Airplane), Commercial Pilot (Airplane), and Flight Instructor (Airplane). We have all the answers and references with analysis of each question: if you are about to take an FAA written exam, these answer books are a must. Write us now to be put on our mailing list for the new books as they are published.

BOOKS NOW AVAILABLE
Private Pilot Airplane Answer Book.

The cost is listed as $6.95.

And there, indeed, is your "pole" if you are inclined to "vault" right over the examination mountain. It really doesn't take many hours of study to get both the questions and the answers pretty well in mind; it can become simply a matter of rote drilling. In this respect, the written examination is a sham. This is especially true when you consider that a passing score is seventy percent, which is really pretty low.

On the actual exam, the only change in the booklet that the FAA made was that the questions started with number 201 instead of number 1 and did switch around the order of the alternative answers choices available for the objective questions; but if you could remember the answer as well as recollect the question, you simply searched out the answer since neither the questions nor the answers themselves had changed.

I would feel sorry for anyone who approached the examination using the rote memorization method. They're missing the enjoyment of really learning about flying, not to mention that ignorance is a dangerous qualification for flying. How can anyone really feel that they have accomplished anything or can feel comfortable in terms of their abilities and knowledge with such preparation? But the point must honestly be made that it can be done in that fashion and undoubtedly is done in precisely that way.

Another advertisement (from the same issue of the same magazine) points this out by stating:

> Your license is worth more than 600 questions. For a couple of dollars the FAA will sell you a copy of the 600 questions it has for each pilot written exam. Even the answers are up for sale. And you could memorize them all. But you still wouldn't be a qualified pilot unless you knew the subject matter.

They then go on to advertise their particular flight manuals.

An alternative to memorization is attendance at an abbreviated, accelerated course of instruction designed to prepare you for the test. There are a number of these "schools" that have mushroomed in the last few years, charging between $75 and $150 for a weekend of study where they "guarantee" that on the following Monday you can take the test and pass it. The guarantee sometimes takes the form of the right to retake the course if you don't pass the test; some simply give you your money back. They risk very little, I am certain, with the average individual since they too know what the universe of questions will be and simply work hard to pound across in some fifteen or eighteen hours, usually spread over two days, enough in the way of information, questions, and answers to get just about anyone through the exam. There is also no question that the students who do attend these "schools" do pass the examination for the Private Pilot's License with little difficulty.

Another alternative is to take a longer route and attend a ground school as offered by certain groups or organizations. For example, many high schools offer a Private Pilot's Ground School Course in their adult education division and these are usually very well run, continuing over a number of weeks. Sometimes a private company having a major operation at a

Beech Student pilot kit

local airport will offer such a course. These courses must be contrasted to the kind of ground school training offered by Cessna, Piper, and the other aircraft manufacturers who are in the business of promoting flight training. This latter kind of ground school is integrated right into the flight training in that you participate in ground school training (reading the text, listening to tapes, doing problems, taking quizzes) as you are learning to fly.

Regardless of what technique you use, you have to get some kind of "diploma" or certification indicating that you have completed ground school from someone authorized to issue such a document before you can take the examination. My own experience has been that the demands of the Cessna course were far and above more difficult and much more all-encompassing than that required for the FAA examination. The Cessna course had exercises that were harder and required a great deal more in the way of knowledge and effort in order to successfully complete the course, pass the quizzes and

Cessna pilot course kit

successfully complete the sixty-question final examination. I am confident that is true also of the other major manufacturers and their courses. Anyone who can get through one of these courses and successfully complete the required material has got to pass that FAA examination, though it, of course, would not hurt doing what I did, namely, having completed the Cessna course, bought the book with the six-hundred questions so I could study for the examination. What became evident after buying the test book was that the FAA questions were a lot easier than the questions in the Cessna course. Furthermore, the private tutoring, the tips, and explanations I got from Tom were more than adequate for handling the examination process.

I set my sights for a particular midweek morning and showed up promptly at eight o'clock in order to complete the examination and get down to work. As I had calculated, I was the first examinee, and it wasn't until I was almost finished that a couple of others came in to take their examinations. You are allowed four hours to complete the examination and the average examinee completes it in about two and a half. I finished the exam in one and a half hours; the person to whom I returned the examination materials was rather surprised at my speed. I wondered why anyone would take much longer (although I had really rushed) given the fact that you knew what the questions were going to be.

As I was told it would happen, about a week later I had the computer read-out sheet from the FAA indicating I had achieved a score of 88 percent; I was somewhat disappointed that I had missed more than I thought I had, but then it was obvious that I rushed the matter and committed some stupid blunders. Nevertheless, 88 percent was a comfortable margin in terms of passing the examination. Tom later told me that this was the exact score he had achieved.

7

Dual Cross-Country Flying

How far I had traveled since I had first picked up the large Cessna manual and began to read about flying! Again I would analogize to that which is experienced by most law students; it's amazing how much you learn in so short a period of time. Before me now was the flight test. It seemed a long way off given my slow progress in the air. The average student gets his Private Pilot's license in 66 hours of flight time; I was coming close to that figure and was nowhere near ready for the flight test.

Though I firmly believed the best route was to stick to one instructor, not to gypsy, the impact of bad weather (cancelled lessons with Tom) combined with a break ("Yes we are flying. Rick is available") impelled me to head out to Midway for a lesson with the chief pilot of the Flight Center. Rick Bodee doesn't look much different from Tom (Are there any really overweight flying instructors? They can't be too big or they wouldn't fit into the small confines of the cockpit) nor was he far apart in age, but they were certainly different as instructors.

Rick only vaguely knew of my status as a student; he had seen me around, apparently knew my profession, and probably knew I was slower than the usual student in my learning aptitudes. I wondered whether the instructors talked much about their students; of course they must. After all, there was no professional obligation in the sense of fiduciary confidence, such as attorneys and physicians must maintain.

81

Rick immediately informed me that none of the 152s were available, not because they were up flying, but precisely because they were *not* flying—they wouldn't start. This Chicago winter was cold (they all are) and snow abounded (a record heavy fall!) creating every sort of cold-weather problem. Cessna had switched to a 28-volt electrical system (from 14 volts) to get a stronger "shot" into the starting system, but to no avail. The battery of an aircraft of this type is much smaller and lighter than that of an automobile and just doesn't have the punch. Furthermore, with the 14-volt system, it was easy (and done frequently in winter) to jump the batery using a car battery; but with the switch to a 28-volt system, this jumping process just wasn't practicable.

Winter is really rough on flying and airplanes. The movement of cold fronts, snow, and cold temperatures combine to insure little in the way of predictable starting or flying conditions. What about hangers? Great, a heated (even unheated) hanger insures starting, but around Midway the cost appears to be prohibitive. My Flight School had none and from what I gathered, neither did any of the competition.

Had I flown a 172 (Cessna Skyhawk) before, Rick asked. I responded, "twice," but immediately excused one lesson as "very brief;" the other, as already noted, was less than stellar. The differences between Rick and Tom were immediately apparent and would become more so through this lesson. Rick was more forceful and more plunging in his actions and attitudes about planes and flying; they damned well better do what he wanted to do. He kicked the tire with vigor when one brake froze on the plane we were to fly, evincing gusto as well as purpose in the act. Yet, there was a softness in his demands on me on the level of, if not excusing, then at least, "its OK," to a greater extent than did Tom. Perhaps it was because I was someone else's pupil and this was just a "fill in" for his time sheet but somehow I got the impression that this was indeed the style of the man.

The engine started; given the cold weather, I was thrilled. The 172 came to life and roared beautifully. It should be kept in mind that while we give little heed to the working parts of our cars (if it starts and runs in winter, that's it!) with aircraft they must be in as good shape as at any other time; everything

Cessna Stationair 206

must be right. That goes for the engine as well as all operating and non-operating parts. How many times have you (and I) for example, failed to clean off more than a little of the windows of our car when covered with snow? Well, you can't do that with airplanes; they have to be clean or your wing loses lift, and snow or ice create other dangerous conditions. So, out with the brushes and brooms and off with all snow and ice.

If the panel containing the instruments of the 152 is complex to the trainee, the 172 is far more so; not only are there many more instruments, buttons, and dials, but, as noted, their position is different. It was evident, however, that both safety and convenience were built into this Skyhawk. Two radios and two sets of radio navigation aids, an Automatic Direction Finder and a Transponder yielded these advantages. There was backup if one system failed together with greater navigational facility and the benefits of expensive, refined instrumentation.

Rick had me fly out of Midway, which I performed smoothly and then used the Automatic Direction Finder, a new experience for me, to "home in" on a commercial radio station on the way to Frankfort airport. On the way there he had me do what is known as minimal controllable airspeed, involving slowing the plane to landing airspeed with full flaps extended and slowly turning in different directions. What a curious and really pleasant sensation. The plane barely moves; the earth and sky seem suspended in slow motion. Rick indicated he was satisfied with my work.

We found Frankfort, which was not an easy task. What with high drifts of snow from the last snowstorm, finding that single runway (most of which was snow covered) and coming into a pattern for touch-and-go's demanded intense concentration. To my surprise and delight, my landings were quite good and thus quite different from my last Skyhawk flying. I was still coming in a little fast, not flaring out soon enough, but all the touch-and-go's were decent enough. Furthermore, I got back into Midway with no problem.

What I had learned from Rick was that mastery over the plane was more important than "doing it by the book;" if it took more airspeed or less throttle than that stated in the specifications, do it that way, the way that gets what you want from the plane. Rick spoke rapidly, corrected little, and basically wanted only the final results to be there; how you got there was less important. Tom was obviously more of a "by the book" instructor. To Tom, how you got there was as important as the final result. Tom was also, apparently, more cautious and safety conscious. I got an illustration of that when, during the slow-flight maneuvers, Rick suddenly ripped off his wool hat, smashed it against the instrument panel covering a few of those instruments and told me to control and maneuver the plane. Tom would have *told* me not to "chase the dials;" Rick covered them.

On the way back into Midway, I received the tower instructions. Rick asked me whether I had understood them. I had and proceeded into a wide circle so as to be on a downwind for Runway 31. Rick reacted with some ambivalence; "Why don't you tell them a base would be more direct for you; oh, forget it—you understand what they want, but sometimes you have to talk back." I was, on the other hand, entirely pleased that I could understand and perceive what was necessary to get into the pattern for a landing, on the right runway and, furthermore, that I wasn't terrorized by the thought of flying into Midway.

It had been a good lesson and tremendous fun, all the more so because it was done in the Skyhawk, the aircraft I expected to fly on a regular basis once I got my license, while the last time up had seemed to be more than I could handle. But a development of the next two weeks changed the scene—all training would now be in Skyhawks. The reason: weather and something of a minor revolt by the pilot-instructors.

Flight instructors get paid when they fly and thus are dependent on aircraft being in good supply and operating when you want them. Neither of these prevailed at the Pilot Center and the flight instructors were fed up with 152s that wouldn't start or had some defect. They were convinced that students would rather pay the $2.00 an hour more than the 152 cost (true!) and could be convinced to take their instruction in the 172 Skyhawks (also true!). Tom questioned me as to my willingness to accept a still further raise in rates above the Skyhawk cost which he said would allow the operation to put one or two planes on the line for student instruction and rental use. My response was in the affirmative; the small additional investment would, I hoped, relieve some of the frustration over cancelled lessons and unavailability of aircraft. The rate was raised to $25.00 an hour for the 172 and $12.00 an hour for dual instruction. Fuel was included in the cost.

The change to the Cessna Skyhawk meant some important changes for me and those who, like me, were in the learning process. Little things, like buying a new operators manual; the one supplied with the Cessna kit was for the 152. Major things, like "psyching in" on a new airplane, not as a substitute plane when, as had happened to me, the smaller 152 trainer was not available. The difference in the planes hit me the first time I got into the cockpit. "Why this is a Cadillac compared to a Honda," I exclaimed to Tom. It seemed no exaggeration. It was so much larger, so much better appointed, more substantial and, as noted, more complex than the smaller, standard trainer. This then was the plane in which I would complete my training and use for the flight test.

My first "official" 172 flight was with Tom on one of those rare interludes between the snowstorms and extreme cold of the Chicago winter. This was to be an intensive session because, with the commitment to the 172, not only was there a new aircraft to learn, but I had to quickly gain the proficiency necessary to solo so as to keep up with my progress toward the license goal. While I had been certified as competent to solo in the 152, each different make and model of aircraft required a separate certification by the instructor. So, off we went to Frankfort airport, this time by pilotage, backed up by radio navigation, the latter mainly to see a demonstration of its use in figuring where a point was off a particular radial from a VOR transmitter station. The session was not very promising. It seemed that I was back to "square one," making so many of the same errors in plane handling and in the touch-and-go's. But, though I ended the session on the "down," I was determined and returned the next day.

Two solid hours of touch-and-go's at Gary and Tom felt I was ready for soloing the Skyhawk. This session had been what it was designed to be, grueling and grinding, in order to build decent handling proficiency. I did, at last, feel that the plane was becoming "mine" and gratefully (as well as proudly) presented my student license for Tom's endorsement before he stepped out at the entrance to the Gary Control Tower. I did the usual three times around the pattern. The first landing was fair, the second excellent, the third rather poor (I bounced). Taxiing over to pick up Tom. I wondered what his reaction would be. He reported seeing only the third (and worst) of the landings! What luck. That left me on the defensive, explaining that the others had been better. On the way back to Midway, we practiced slow flight and then back to home base. While much more work was necessary with the Skyhawk, Tom felt that we should now do that first cross-country dual flight, the one that had been planned for in the 152 some weeks before, but then the weather and a change of planes and plans had intervened.

It may well be asked, why all this up and down business with the plane, this pattern work; doesn't it get boring? To answer this second part first, *nothing* about learning to fly can be labeled boring; tiring, yes, but not dull. There is simply too much challenge and too much at stake in the learning process. The first question is harder to answer unless you have flown, and that flying includes (because the same holds true) radio-controlled aircraft.

Pattern work, that is, taking off, going crosswind, downwind, turning base and coming in for a final approach, touching the wheels and then up and around again, in effect flying a rectangular pattern around the runway in use, is *the* basic flying exercise. Essentially flying a pattern incorporates learning, developing, and perfecting many all important flying skills. You start with the proper take-off technique, that is, when to "rotate" or lift the plane off, based on your developing airspeed as well as your feeling that the plane has developed sufficient lift. You must keep the runway heading (not wander off to the left or right) as you ascend at the proper angle. Then, when you have left the complete runway behind and have traveled far enough beyond it, you turn left (while you're still climbing) into the crosswind leg. The amount of crosswind you will travel will vary with the many factors, some of which are pretty subjective and judgmental. Then left turn again (you're in the normal left pattern) and you have come to your proper pattern altitude (800 feet plus the field elevation above sea level) and you level out, commencing preparation for landing. Carburator heat is pulled on when you see that you have come halfway down the runway; drop the engine to 1500 RPM when you have reached the end of the runway and go into slow flight; adjust the elevator trim; get the airspeed into the "white section"; put in 10 degrees of flaps and slow to 60 knots per hour. Turn base, add more flaps; turn final, maintaining slow-flight airspeed right on the button; develop the proper angle of descent, then round off above the runway and flare out to touchdown.

Taken together, you have all of the elements of flying except long-distance navigation, that is, the takeoff, climbing to a set altitude, maintaining a course, setting up for landing through slow flight and the landing—it's all there in pattern practice. Indeed, in flying radio-controlled models, you soon learn that it takes little practice to be able to simply fly around the field without losing control of the model; it is in the pattern practice that the pros are separated from the amateurs. Ultimately, the climactic point of the pattern is that touchdown, from which power is applied and around you go again. It has to be right; right down the middle of the runway, with the wheels touching the pavement (just "kissing" on the rear wheels if you've done it right) in the first third of the runway. Furthermore, if all of this is done at an airport with a control tower (such as I have described at Gary, Indiana) there is the added demand of working with the tower on the radio as the pattern is entered, worked, and then departed. With the control tower comes dealing with other aircraft entering the pattern for landing, or those taking off from that field, or simply passing through the control area or other trainees doing touch-and-go's. It's an additional demand, keeping in continual communication, as well as adjusting your flying to that of other planes using the field.

The licensing requirements say you must have logged a minimum of twenty hours of solo time, with at least ten hours of solo cross-country. With this in mind, Tom began emphasizing more solo time, both to meet the requirement and to build confidence. The latter, that all important confidence begins to come through with solo time, no doubt about it. You really begin to feel that you are controlling that plane, not vice versa, as the solo hours mount in your log; it's a good feeling to know. It's also time for cross-country practice—dual and then solo.

Our first cross-country flight (one more than fifty nautical miles from takeoff) was planned for Dwight, Illinois, southwest of Chicago. The planning brings together what you have learned in ground school with the flying of the plane to a specific destination in conformity with that plan. Tom sat across from me at a desk while I got out the sectional chart, drew a line (the true course) between the airports (Midway to Dwight) and made the necessary calculations including figuring the altitude for the flight, the time it will take, the fuel to be used, and the heading to be flown. The heading indicates the true course corrected for wind and other factors. Forms are used for this planning, but the linchpin of it all is weather. So it begins with a call to the government weather office by calling an unlisted number reserved for pilot use, identifying yourself by the number of the plane you will be flying, and asking for the pertinent weather (both present and forecast) along the route of your flight. Have your pencil ready; they give it to you (ceilings, visibilities, surface winds, winds aloft, forecasts) in rapid-fire succession. This weather briefing (which can be in person or by phone) helps you decide whether you can make the flight safely under Visual Flight Rules (non-instrument flying) as well as garnering the information necessary to complete your flight plan. For example, the true course from airport to airport may be on a heading of 220 degrees but at the altitude you will be flying, given the wind and other variables to be taken into account, you may determine (using your calculator) that the heading to be followed to get your destination will be 230 degrees.

Along the route you also tick off approximately every fifteen miles on the map in order to have a visual reference point on the ground so as to insure you are on course. This could be a major road, oil storage tanks, a town, railroad tracks, a lake, or similar geographic or man-made object or distinct terrain feature that should be seen below or to one side of your course as you fly. This kind of flying combines what is called pilotage (flying from a seen point to the next point to be observed) with dead (deduced) reckoning, which involves calculating the time to elapse from point to point under anticipated conditions of flight, including wind, ground speed and the like. You learn these methods even though most long-distance flying is now done by radio navigation, which allows you to fly from one place to another by following a radio signal displayed on an instrument in the cockpit. Your training includes learning various radio navigation systems, but your cross-country-flight training puts heavy stress on the older visual systems of pilotage and dead reckoning. Along the way, you can't help but begin to appreciate what air navigation was before radio navigation aids were perfected. If the air pioneers of the years up to the Second World War are most often cast in the mold of fearless adventurers who survived on a diet of luck, guts, and instinct, well, just imagining what they had in the way of assists (practically none) to help them know where they were in the sky and where they were heading, leaves any pilot with his hat off and thankful for all the electronic marvels that make sense out of the sky.

Normally, a flight plan is filled with the local Flight Service Station, a system of offices maintained by the government to give in-flight assistance of a variety of sorts. We didn't file the flight plan I had prepared because the flight was a short one (thirty-seven minutes as planned) and the weather seemed to present no problem. Had a flight plan been filed and we did not show up at the listed destination a search would have been initiated. These stations, among other services and activities, provide a constant monitoring of the emergency frequencies; it's comforting to know that somebody is listening for your distress call and watching out if you are late in arriving.

The trip to Dwight was sheer fun. The map came "alive" as I picked out the checkpoints as well as noting all sorts of geographic details. How different it is from driving along a road with the limited vision available from the ground. Just watching how roads curved, a thin line of blue river appeared, and towns came into sight made this flying exciting. Of course, the objective was to know exactly where you were in relation to these ground features, but I found what so many before me had discovered in flying, a feeling of freedom from the hold of gravity, the joy of experiencing the dimension of life called flying. And it's not the same in an airliner; generally you fly so high and so fast for most of the journey that this kind of relationship to earth, yet freedom from it, is neither witnessed nor experienced by passengers.

I managed, however, to fly right by the Dwight airport. I realized I had gone too far, noting certain geographic land features against my map and turned around in a slow circle. But it was no use, I just couldn't see that little airport and its one runway. Tom pointed it out to me, remarking that as I turned, by chance, I had kept a wing tip just over the airstrip, blocking my sight of the field. We went into a pattern over the airport and then shot a couple of touch-and-go's as well as a full-stop landing. These were not very well executed; I had a tendency to let the plane wander off the center of the small runway. Tom made the point that such inaccuracy wasn't too serious when the surrounding ground was frozen hard, but in other seasons it might well be soft and muddy if you ran off the strip of pavement and found yourself mud-bound.

On the trip back to Midway, Tom had me practice using radio navigation and it went well. I had no difficulty finding Midway or getting into the direct landing position. What a welcome sight that hugh runway was, yet the trip to Dwight made it clear that the right procedure was to land with the nose right down the *middle* of that runway. The first dual cross-country was complete. Ahead lay a more lengthy dual trip, solo cross-country work, and lots of solo practicing.

There was a special, a personal, meaning to asking Tom

whether I could plan our next (and much longer) cross-country to Champaign, Illinois. My daughter was there as a freshman student at the huge University of Illinois campus; for the first time I would be flying somewhere "meaningful." Furthermore, it would be a route I would hope to do often, just to "drop in" or take her home or back for the weekend. Now that's putting a practical side to the entire scene! So, with Tom's concurrence, I planned the trip, laying out the true course, making the corrections for reported winds aloft and calculating the usual checkpoints. But then, Chicago weather took care of these careful plans—three times.

In the interim, I had occasion to drive the 125-mile trip to the University with my wife for a one-day visit to see Karen. The weather at first seemed frustratingly good for flying; what a waste to drive. As we went along Interstate 57, however, we experienced a strange weather phenomenon. While the sun seemed to shine brightly, the visibility was poor, incredibly poor. It wasn't fog, exactly; it was low lying clouds. No, this was a driving, not flying day. Once again the important limitation for VFR (Visual Flight Rules) flying was evident; you can't fly safely in this kind of low visibility weather, which, from the looks of it, could fool the unwary into initially believing all was well.

The road to Champaign parallels the Illinois Central Railway tracks and constitutes as nearly a direct route as can be flown; put another way, running a true course on a map from airport to airport, you are seldom out of sight of either the highway or railroad and most often, both are in view. So, if I couldn't fly it that day, I would "mentally" fly the course by tracking the route as I drove.

Fortunately, the road was only lightly traveled, so I could concentrate on the sectional air map as well as the traffic (a scene my wife was less than thrilled with, but then she is somewhat used to seeing me do more than one thing at a time). Doing more than one thing at a time is in fact a needed talent in flying, perhaps more accurately, doing one thing at a time, but moving very quickly to the next and next and next. So, for example, you are

• keeping the craft in trim for straight and level flight
• listening to the radio

- keeping your course
- maintaining your altitude
- watching for other planes (a constant demand)
- and navigating by observing instruments and visual references.

Thus, it not only seemed a lot less demanding to drive the course and watch my air map than fly it—it was.

An interesting footnote to this trip: It seems that a pilot in a Cessna 172 found his plane icing up in the very region we passed that day and landed his plane safely (and with great skill) on the road leading to the weighing station from the main Interstate 57. When the weather cleared, state troopers cleared the traffic on the main highway and he took off!

The first fact that impressed me as we got under way by plane when the weather finally coincided with a scheduled session was that the "rule book," the FAA regulations and procedures, were coming alive. Here I was, really calling in my flight plan, maintaining less than 3000 feet on the altimeter until beyond the Terminal Control Area on the map, talking to the Flight Service Station on the radio, monitoring Flight Watch (inflight weather service), climbing and maintaining the prescribed (and calculated) altitude for a VFR flight in that direction, as well as doing what I had done on that dual flight to Dwight—watch for the visual checkpoints. The 105 nautical-mile trip just zipped by; not another plane in sight. Soon I was passing near the old Rantoul Air Force Base (now closed) and at that point, in retrospect, failed to realize I should have begun descending for entrance into the landing pattern altitude (1600 feet). What stupidity; it can't be put any other way. Like chess (I never could play the game) you have to think far ahead, just as you have to look far ahead as you fly, not down or just around. Did I think I would suddenly go down to a landing from 3500 feet? When I realized my error (Tom said nothing) I was already in sight of the field, so that, in addition to getting landing instructions, I had to get down and do it too quickly. One error lead to another; I told the tower I was south of the field when I was north of it. Finally getting them (and myself) straight, I landed with less than great style.

It became quickly apparent that this airport was incredibly busy. Two planes were doing touch-and-go's on one runway and other planes, including me, were landing on another. I felt as though I was landing at O'Hare, certainly not Midway since I had never heard nor seen so many aircraft nor so much traffic.

Over a brief lunch, Tom explained that this field was used by the University of Illinois School of Aeronautics and there were students doing flight practice. Sure enough, they came piling in to the cafeteria for lunch, dozens of would-be pilots and associated flight trainees. It has always been of keen interest to me to note the kinds of people in flight training. These were all young, college-age students, not unlike the variety of students with which I had spent my adult life, except that they were in a flight-training program, not in law or history studies. And here I sat among them, old enough to be their father and seeking to learn to fly, not for career purposes (which was their aim) but, nevertheless, striving to gain some of those same skills. No doubt about it, learning at their age is easier and quicker; middle-age abilities are not those of a twenty year old. Yet many of them would not make it. The flunk-out rate for freshmen in the School of Aeronautics at the University is higher than any other school.

With fuel tanks topped off and a second try at reaching my daughter (unsuccessfully; she was in class) the return flight commenced. Once again, the airport was buzzing with action. As we were about to take off, a twin-engined plane landed on what appeared to be a taxiway; Tom recognized this (I didn't) but couldn't explain it. The next time I met with him, he told me he had "asked around" and, indeed, the airport was so busy at times that skilled pilots were allowed or even directed to land on the taxiways. On the return route, I was to practice radio navigation, "chasing the needle" as it is called.

After the Second World War, a new system of navigation was developed, replacing an earlier, less reliable and more difficult system. Today this system, the radio-airway system, is the basic navigation tool for all commercial, military and, private aviation. If you have wondered why the windows in the cockpit of an airliner seem so small (and they generally are) and questioned how the pilots could 'see where they are going," the answer lies in the fact that, for the most part, they don't "see" in the usual sense, in order to navigate in the air. Scattered around the world are radio beacon devices that generate radio signals which can be used to guide pilots from one point to another with great accuracy. Once reaching the vicinity of an airport and assuming a visual approach (seeing the airport and its runways) is impossible or the attempt to do so inadvisable, the plane is guided by the work of the men in the Control Tower. Separating the planes from each other's paths, in combination with various instrument landing systems, enables the pilots to guide the plane in safely in almost any weather. This air navigation system is an unsung modern miracle of aviation science, quite complex and yet unfailing, with lots of backup systems as well. We read of the tragedy of airliners crashing from time to time, but not of getting lost—they just don't.

Even small, single-engine planes can be (and today generally are) equipped with enough of the necessary radio-guidance instruments to take advantage of this air navigation system. Their cost is high, running into thousands of dollars and frequently amounting to between one-fourth to one-third the cost of the aircraft itself. The plane I was flying was so equipped, and thus the trip back was to be based on practice using this radio navigation system.

The trick is to set a receiver in the plane to the signal of a transmitting station and then track to that station following a visual indication that takes the form of a needle which centers when you are "on the beam." "Trick" is too easy a word; "task" is better because knowing what you are doing is all-important and gaining skill in tracking the proper course requires practice. For me, the tracking went fairly well, but the "idea" of how much course change was required to recenter the needle seemed illusive.

This VOR work, as this radio navigation system is called, took me back to Midway, or at least it would have, had I not goofed and lost the conceptualization of what I had to do after passing over the last VOR transmitting beacon station. All I had to do was keep flying north and I would have run into Midway; instead I veered off-course and started visually hunting for the ground clues that would help identify where I was. Everything mentally seemed to come apart. The strange thing was that I knew it, that is, I knew I was mentally confused and ruining my otherwise acceptable performance in this dual cross-country. I was angry (and embarrassed) with myself, even though I did, after stumbling around, get into the Midway pattern and performed a good landing. Tom was silent; I was glum.

When we got back to the office, Tom took me into the student study room and with chalk in hand, described what I had done wrong. To me, it wasn't the error, but the near panic that hit my mind was what troubled me. Not that my navigation goof was of no consequence—it was—but there was no need to become so disoriented, which is probably a more appropriate term.

The lessons were clear; know more about what you are doing, and when faced with a problem seek the alternatives available. Rationally, I knew I couldn't be that far off. Nothing "dangerous" had developed. With plenty of fuel, decent weather, good control of the aircraft, and radio communication as a means of obtaining help, there was no real need for concern. In some respects, I was simply angry with myself for having carried too much of the world of driving into flying. No,

Home study

you can't stop, pull off the road and figure things out. No, there are no "gas stations" in the sky. But you do have that map, your instruments, and the radio. You are also taught the four "C's" when you need help:

Confess—that you are lost or have a situation you can't handle.
Communicate—with a control tower or flight service station about your problem or if it's a real emergency, get on to the emergency frequency which is constantly monitored.
Climb—for better reception of radio signals.
Comply—with what you are told to do.

You are also taught that, unlike the driving situation, altitude (get plenty of distance between you and the ground) and speed (keep up your airspeed) are your safety factors in flying; low and slow are the wrong combination.

Furthermore, just as in any endeavor, you have to face the reality of hitting a plateau in the learning cycle—a level of proficiency is attained and then you struggle to break out of that plateau and get climbing once again, improving your skills and acquiring new ones. I had certainly faced that phenomenon in learning to fly radio-controlled models and the same is obviously true in every sport as well as any vocation. *The* important plateau for most aviators is landing (true of models and full-scale); I had "climbed" that one, but now I had to work on this navigation business which involved breaking earthbound habits as well as developing that which I simply seemed to lack on the ground or in the air, a sense of direction and confidence in handling temporary disorientation. The willingness to face the hard work involved and overcome your particular handicaps is what it apparently takes.

8

Solo Cross-Country Flying

Despite the weak performance on that dual cross-country to Champaign, Illinois, Tom immediately instructed me to plan my first solo cross-country. Meanwhile, he wanted me to go up for solo flying, practicing touch-and-go's and building confidence along with the required solo hours. "Where to," I asked. "Bloomington; its a good long trip and its got a control tower," was the response.

Out came the sectional chart as I thought about what I knew of Bloomington, Illinois. Like Champaign, it's a college town, though without a School of Aeronautics, and like Champaign, it's southwest of Chicago, somewhat further west and about the same distance (100 miles) south. The planning went well; I prepared the route on the sectional and made a little diagram of the airport runways so that I could easily visualize them, pasting the diagram near the airport symbol on the map. The weather cooperated beautifully this time and though I was delayed in getting to Midway by an administrative hearing before the Illinois Department of Labor that ran longer than I had anticipated, there was plenty of time for the trip. Tom was there when I arrived and I immediately sensed reticence; "Art, the weather sounds bad down south—I don't know. Did you get your weather briefing?" I explained that I had and that the briefer at the Weather Service Office had stated that the line of showers and bad weather was a little way south of Bloomington and was not expected to move north. Bloomington was reporting a ceiling of 6500 broken and seven miles visibility (minimums for VFR flight are a ceiling of 1000 feet and three miles visibility). I said I'd call again (my last call had been three hours before) and Tom stood over me as I wrote down the latest figures; nothing had changed. Tom took the phone and spoke with the briefer, apparently wanting to hear the report himself as well as to ask some further questions. That kind of concern, genuine, professional and evincing a personal involvement were deeply appreciated and not at all unique. "Well, it looks OK," he stated, "but remember . . . " I cut in on him with, "I know, do the 180 if it looks at all bad." The "180" referred to is the expression of the key safety factor for VFR pilots (non-instrument rated) and simply denotes the command to turn around if the skies ahead seem to contain threatening weather. Failing to perform the 180 at all or early enough is what has been responsible for most weather-related accidents involving VFR pilots.

Having checked my flight planning work, Tom gave me some last-minute instructions and I drove off to the waiting Cessna 172—numbered 734DB. There is that certain feeling, captured and recaptured in writing about flying since the Wright brothers time, that feeling of "its just me and the machine;" that feeling, not at all unpleasant, was with me as I adjusted my seat height, scanned the instruments, and snapped the seat belt together on the seat next to me so that it wouldn't flop around, a clear symbol that indeed, it was just me and this aircraft. It's a feeling of aloneness and yet of being a part of a highly populated, though different world—the world of flight and aviators.

My pre-flight completed (checking the moving surfaces, the oil, and all items on the checklist), I received permission to taxi to the outfit that supplies gas to my flight school. I polished up the windows (they're plastic) as best I could, rechecked the pre-flight checklist and took a long last earthbound look at that sectional chart.

Within a matter of minutes in the air, I had established my climb and heading (direction) and was busy looking for the first checkpoint. It came quickly by and just about that time I was able to climb to the cruising altitude. I called the Flight Service Station saying, "Chicago Radio (that means you are calling the Flight Service Station for the Chicago area) this is Cessna

734DB listening on 122.5 (one of their listed frequencies)." In a moment, a woman's voice came back responding, "Cessna 734DB, this is Chicago Radio, over." "Request opening of filed flight plan; 34DB." Pause. "We have it and opened, 34DB; have a good day." With that bit of pleasantry, I was now in the "system" so that if I didn't report closing that flight plan within one-half hour from when I said I would be back to Midway (I filed a "round-robbin" so as not to have to refile at Bloomington) a search would be initiated. Once again the map came "alive" below me; there was a river, there oil tanks; way over there a superhighway, over to the left a small town; the world from the aviator's view—so orderly, so manageable, so defineable.

I was a little concerned with two aspects of this flight. First, I felt that the directional gyro—the basic direction or heading indicator—was off. You set it to conform to the compass, but sometimes it wanders off, or the compass, is off. The compass (in this aircraft, hanging from the top of the windscreen) itself is not used for navigation because it bounces around too much. Normally, you don't have to check on the need to adjust the directional gyro more than once every fifteen minutes, but here is was a little less than that and it seemed off. I adjusted it to conform to the compass reading as best I could, though there still seemed to be something wrong. Then it came to me; while the first checkpoint and then, coming up, the second, were coming into sight and clearly recognized, they were (meaning I was) on the wrong side—to the west of where they should have been given my course line. So I banked over to the east and got back on the heading, this time with the correct orientation— that which should have been to the left or right appeared in proper place.

I was struggling with another problem in my mind, simultaneously with the orientation problem. Was I just imaginging it or did I lack power? The airspeed seemed a little slow and so did the RPM's. I shoved the throttle further in and in doing so realized what the cause of this problem was; there is a device for tightening the throttle (the throttle is a kind of plunger that sticks out of the dashboard—push it in and you open it; pull out and you slow to idle) into a pre-selected place. This device was loose enough to allow the throttle control to ease out,

Cessna Skyhawk II

slowing down the engine. I merely tightened it and the power stabilized at the desired setting.

Tuning in Flight Watch, an in-flight weather service, I listened as other pilots in a wide area called in for a weather information update on the locations to which they were heading. Here was one pilot on his way to Fort Wayne, Indiana, (there had been flooding there and someone was trying to get in by air) another to Chicago heading in from St. Louis, still another flying from Chicago to Springfield. When the air was silent, I pressed down on the mike button, saying, "Chicago Flight Watch, this is Cessna 734DB," and waited. A quick response; "Chicago Flight Watch; go ahead Cessna 734DB." "34DB over Joliet Arsenal, heading for Bloomington, VFR; Can you tell me weather and winds at Bloomington?" "34DB, which Bloomington? Illinois or Indiana? Over. Goofed again; of course there are two Bloomingtons I could be going to. "34DB, Bloomington, Illinois. Over." "Ah—weather at Bloomington, Illinois is reporting 7000 feet broken, 11 miles visibility, winds about 060 degrees at 6. You're OK if you stay at Bloomington or north of it; weather is south of the area. Over." "That's it; thanks," I responded. He had anticipated my concern as to whether that precipitation had moved into the Bloomington area; with the good report given, I continued my southwest flight.

I had marked on the chart a point about ten miles north of the airport where I would contact the Bloomington control tower and began my descent; no more goofs here—plan ahead! My call to the tower was as follows:

> Bloomington Tower, this is Cessna 734DB, about ten miles northeast for landing. Be advised, student pilot on cross-country not familiar with your airport. 34DB.

> 734DB, Bloomington Tower. Winds 060 degrees at 6 (knots per hour); altimeter 30.02; report downwind for Runway 3.

> 34DB.

The altimeter setting is given so that you can reset your altimeter to the new setting for barometric pressure at that airport. In turn, that allows you to have the correct altitude measurement shown on your altimeter since the altimeter measures altitude based on barometric pressure. Furthermore, the ground elevation as compared with sea level at Bloomington is about 225 feet higher than at Midway. It's 875 feet, to which you add 800 feet to get the pattern altitude of 1700 feet (rounding out to the high side) that must be used in entering the landing pattern.

The Bloomington airport is north of the city, and, as I descended, it came into view over to my left. Glancing at the little airport diagram I had pasted on the sectional chart, I figured where I would have to be to report on the downwind leg of the pattern. Reaching pattern altitude, I entered that downwind leg, carburator heat on, slowed down to 1500 RPM, 10 degree flaps, turned base, 20 degree flaps, nailed 60 on the airspeed and was ready for final when I realized I was off on the approach, not properly lined up with the runway. I corrected with slow flight into position and landed. The tower had cleared me for landing as I reported on the downwind leg. Switching to the ground-control frequency, I was directed to a parking area; at first going to an area where parking was prohibited and then moving out when ground control told me about it. Almost the moment I shut down the engine, a gas truck arrived to sell me fuel. I waved him away since I had plenty for the return trip. (This aircraft carries about four and one-half hours worth of fuel). As I left the plane, the truck driver offered me a ride to the airport building and I hopped in for a quick ride.

The airport lounges were surprisingly elaborate with comfortable facilities for departing and arriving passengers as well as sporting a very nice restaurant. Quite a change from Champaign, though the airport there seemed much busier. A quick cup and I trotted back to the plane, thinking how different was the feeling of being a pilot leaving the lounge area for *his* plane, rather than playing the passenger role—an interesting and quite pleasant sensation. I wondered whether I "looked" like a pilot.

The return trip was to be by radio navigation only, so I set up the outbound course and mentally planned what I would have to do to get on the selected heading. With lots of help from ground control, I got to the end of Runway 3 and did my run-up to check all systems; I had done a brief pre-flight before starting the taxi run. All systems seemed fine, except that I noticed a few rain drops on the windscreen; had that weather moved north? Visibility was good and the ceiling overcast, yet high.

After takeoff, I established my course heading tracking an outbound radial signal on the radio navigation aid. I was still climbing to the proper flight altitude of 5500 feet when I dialed in Chicago Flight Watch and got a firm indication that so long as I flew north, I would be OK and the weather would improve.

Flying at 5500 feet was the highest altitude I had flown since

I started flight training. Visibility was very good, but I was most impressed by the smoothness of the air, the smoothest I had experienced! Generally, the higher, the smoother and this was pleasant. The land I flew over as I navigated following that radio signal was almost entirely devoid of towns or even any major roads, again more than I had previously experienced. There were landmarks I could pick out: a meandering river, a town way off, but generally, nothing but farms and dirt roads. The matter of keeping the needle centered was my main concern and it went smoothly. I tuned in the second radio navigation receiver to the radio beacon to which I was heading, picked up the inbound radial and flew to it. Once again the map came "alive" as I neared the transmitting station with towns, highways, and prominent landmarks; but these were not my primary guide, that radio beacon was. I could, however, tell from those landmarks that I was very close to it and began to be more vigilant for other aircraft since many aircraft converge at the transmitter, tracking in from different directions.

And then, there it was; I could see the cone-shaped building, the VOR transmitter, sitting in the middle of a field. It isn't necessary to see it for correct navigation, but it told me I was right "on the beam" (in a very real sense) to be able to fly right to and over it. Banking to the left and heading almost due north, I began to plan my descent, but apparently not soon enough since I found I had to go down more rapidly than would be comfortable for passengers not used to light-plane travel. But I knew where I was, what I had to do and got into Midway without difficulty. After alighting from the plane, I sensed that certain calmness. I was secure on land, the challenge was over, the flight complete. On the other hand, the thrill and satisfaction of a good flight, of having met the challenges involved, of making some errors, but none crucial, was a great feeling. Tom was out flying when I arrived back and so I had to share this excitement by phone that afternoon. And it was exciting for Tom too and obviously gratifying that all had gone so well.

Nothing like an uneventful flight to build a form of confidence, the kind that tells you learning to fly becomes less and less a mystery and more a matter of practice. That's how the second cross-country solo went; but then, you also learn that no flying is really "routine" and that you never have a

Beechcraft Sport 150 and Sundowner 180

"perfect" performance. There is always something you forget to do or check or some aspect of your performance that is less than "stellar." The point seems to be to get the errors down to manageable numbers and non-critical ones. Like forgetting to turn on the transponder (not critical) or making a less than great approach to your destination field. I did both on my second cross-country and still felt my confidence deservingly increased.

No doubt about it, the controller in the Champaign tower was less than pleased with my slow, very slow approach to the field and he told the two planes behind me that "the Cessna was on an S-L-O-W final." My ego felt somewhat crushed, but I didn't have the nerve to tell him that I was "peddling as fast as I could." No wisecracks from student pilots allowed.

That second cross-country now logged in my book, I cornered Tom for instructions on flying the third—the long one, where each of the three legs have to be at least 100 nautical miles apart. Throwing out a couple of suggested routes, he seemed to favor Midway to Galesburg, Illinois, Galesburg to Champaign, Illinois, and Champaign home. Since I had done the Champaign route, only the first two legs would be new. So off I went to my task of preparing the flight plan for this final student cross-country.

I worked up pilotage combined with reckoning for the trip up to the point where I would have to have the actual flight conditions, namely, winds and waited. And waited. Cancellations because of weather together with the press of work prevented the flight. The route got stale in my mind as the weeks passed; the weather did allow me a couple of local solo flights, but it had to be good over a wide area before I could go on a day when my work allowed me to go. Additionally, Tom advised me to fly the 152s again, except for this cross-country and take my test in one for a number of reasons, boiling down to "it's easier." He said, "In a 152, you are 'tighter,' cramped alright, but also you have a better feeling, in terms of being part of the plane." It had been three months since he had "signed me off" in my pilot's log as competent to fly a 152, so we went up again in a brand new one just acquired by the Flight Center; this was one of those two times when the spring weather was good enough to fly locally, but not cross-country.

It had to be the windiest day I had ever flown in; the night before a plane at Midway had been flipped over, though tied down and it seemed to me that this 152 might just do that with the two of us in it as we rocked and rolled (with no music) just sitting in the plane and going through the pre-flight check. Flying over to the Gary, Indiana, airport, the hard and gusty wind was fortunately right in line with the runway so there was no crosswind component to worry about, but everything else was worrisome as the high winds near the surface seemed to banter playfully with the plane, saying, "bet I can toss you around a bit," and it did. The effects of the wind were dramatic, making the plane appear to almost stand still at full throttle when going into the wind and then zoom away when banked downwind. I did three of four touch-and-go's, all but one too high; it took a re-adjustment of perception to get mentally in tune with the 152 again. Then Tom directed us up to 2500 feet for some slow-flight and stall practice. For some reason the wind at that level was quite calm and we had a good workout on refreshing the technique of various stalls and minimum controllable airspeed, including some new activities such as stalling in a turn. In all, Tom was pleased and affirmed that I should finish my training for the license in the 152. There was, additionally, a pragmatic reason: Cessna had recalled a great many 1976 and '77 Skyhawk 172s because of serious engine problems. The recall was so unusual and of such dimension that it made the New York Times as well as all the aviation magazines. Two of the four owned by the Flight Center had to be returned. It was an interesting feeling to realize I had flown extensively in both those planes. What if . . .

The day that was "right" weather-and-workwise, came gleaming bright, quite cloudless and not windy. I was to meet Tom for review of my flight preparation at 9:00 a.m. Arriving early, another student and I got into a discussion of my planned flight. It occurred to me that this was the first time I had really "chewed the flying rag" with a fellow student; our urban-hurry life didn't seem to allow much cross-communication between students. This one was also a student with Tom, and he expressed the wish that he was as far along as I; my response and not push the process because it was too important and could be push the process because it was too important and could be genuinely pleasureable. Tom checked my work, gave me some final instructions and then signed my log and gave me the keys to 734DB. I was only moments behind Tom and his student as we drove to the West ramp and I thought we might meet at the fueling station, but he was long gone before I got there.

Pilots, it has been often stated, are a superstitious lot, reflecting the tenuousness of their airborne existence—at least in past decades. How much of this lingers today is uncertain but perhaps I should have sensed this was not to be my day for good flying. When 734DB wouldn't start (the battery was about dead) I was incredulous; it was 55 degrees, not 0! What luck. Back I drove, anxious to get the only other Cessna 172, number 80233 before anyone else got it. Another instructor was there to switch keys with me and I raced back to the field. 80233 had been the plane I had flown most often so I actually felt more comfortable with it.

Fueled, pre-flight complete, a last look at the Chart, and 80233 and I were heading southwest, away from Chicago. After a few corrections to line up my heading with the visual orientation of landmarks on the surface, I called the Chicago Flight Service Station to change the number and listed the color of the aircraft from 734DB to 80233 on my flight plan as well as activate same. I had planned to go by pilotage about two-thirds of the way and then use radio navigation to the Galesburg airport since it has a transmitting station right on the airport so I could "home" right to it. That first two-thirds was bumpier than expected but otherwise uneventful and pleasant. I had already discovered that one of my radio navigational radios didn't work, but the plane had two, so I would just use the one that worked, except that when I tried tuning that VOR at Galesburg, I got nothing. What's the matter; was it my other radio or the VOR transmitter? I dialed in the Galesburg control tower and asked whether the VOR was operational; a kindly voiced controller responded, "Negative; its out today. Where are you; can I assist you?" My immediate reaction that this had to be the friendliest sounding controller I had heard was confirmed; and what luck, the VOR had to be out just on this day. But did I need help? It seemed to me that I was slightly off course, further north than I should have been. Noting a little town off to my left, I told the controller I was near Kewanee, Illinois. "Fine," he said, "just follow the double-track railroad and report when you have the town of Galesburg in sight." As I followed the track, I remembered having read about piloting years ago when following railroad tracks to get from place to place was so common that they were nicknamed "iron compasses" by pilots. Not only did I soon sight Galesburg, but with the clear visibility, I could see the airport and so reported. The controller then directed me into a normal left pattern for runway. Two minutes later I was down. Then I did something I had not done before; extraneous communications with air traffic controllers is frowned upon because it takes air time that might be vitally needed by another pilot. I could not however resist thanking him for the offer to assist me and the plain kindness in his voice and manner, so I did.

The requirement is simply to come to a full stop at each airport on the cross-country, so without doing much more than going to the parking ramp, shutting down the engine, and taking another look at the route from Galesburg to Champaign, a distance of 112 nautical miles (about 130 statute miles), I left Galesburg behind. The flight plan was to use radio navigation for this second leg because the straight course was just about on what is known as a Victor Airway, a specific, frequently traveled line of radio navigation. The backup would be by visual checkpoints. With the Galesburg airport behind and climbing to 5500 feet, I dialed in the Peoria VOR and noted the bearing that would take me there. Everything seemed smooth as I tracked to the VOR transmitting station near the Peoria airport, noting a couple of small towns that passed below off to one side of my course. After reaching the point where the station was located and noting that fact on the VOR instrument, I dialed in the Victor Airway outbound radial that would take me direct to Champaign and then intercepted that radial (the radio signal) by altering the course of the plane.

It began to hit me about fifteen minutes later. Something was wrong. What I saw didn't at all match up with the map. Why? I was tracking the course, but where were the landmarks I should have seen? Maybe I should switch to the Champaign VOR and fly to it. I wished that other VOR radio was working, then I could do both. I switched to the Champaign frequency. No luck, it was too far away for reliable use. Back to tracking from the Peoria VOR. But where was I? Trust the instruments, I told myself, as I ⌣ad in the past. But again, there was no way that what I saw was what I was supposed to see.

It came upon me slowly, kind of creeping from the back of

my head forward, forcing a conscious realization up front—you're lost! And there was the clincher, a large airport of to my right that I couldn't identify on the map. Where the hell was I and how did I get lost? I kept flying, turning this way and that, trying to get a proper orientation. The map—get it aligned with your path of flight. Fold it and hold it correctly. No, still no sense. Fifteen minutes now and you don't know where you are. But the VOR says you're tracking right. No, I'm lost. Better tell someone. Call Champaign Tower, and see if you're not too far for them to hear you.

Champaign tower, Cessna 80233, student pilot cross-country on flight from Galesburg to Champaign; I think I'm lost; who should I be talking to?

Cessna 80233, Champaign tower. We have radar approach here; dial in (he gave me the correct frequency) and they may be able to help.

80233 (by repeating my number indicated I heard and would comply).

Champaign Approach, this is Cessna 80233, over.

Cessna 80233, this is Champaign Approach, over.

I'm flying VFR, student pilot, from Galesburg to Champaign, and I'm lost. Can you assist? 233.

233, have you got a transponder aboard? (An instrument for transmitting a special discreet signal that makes picking out a specific airplane easier).

233, affirmative.

233, dial in code 0100. (They were going to use the transponder and their radar to find me.)

233, dialing 0100. (I waited and thought. So this is what it's like to be lost; not comfortable and you feel like a damned fool, but isn't this also part of the adventure, the experience of flying, I had to ask myself and also admit that it was exciting, with an edge of danger. Not quite like sitting behind a desk, eh?)

233, what's your heading? (I responded)

233, make a 90 degree turn to the right. (I complied)

233, switch code 0400. (I did)

233, we can't seem to find you; can you describe what you see out the window. (I saw a water tower and flew down to read the name on it. No luck, no name.)

233, switch to code 0100 again and make a right hand turn of 90 degrees and say your heading. (I did)

233, we still can't find you; how much fuel have you got left. (I told them, one-fourth on one tank, between one-fourth and one-half on the other. I hadn't even thought about that! Where was my training; all those articles I read about "fuel starvation." Can't be up here all day! There were also considerable delays between these communications as Champaign Approach handled its major task, the sequencing of aircraft arriving to its busy airport.)

233, I think we've found you; make a 90 degree turn to the left and report your heading. (I did)

233, We have you. You are (I thought he said) three miles southwest of Champaign. Do you want to go to Champaign or Decatur. We can vector you (give the correct heading) to either.

233, I prefer Champaign.

233, Take up a heading of 060 degrees and dial in Champaign tower; we'll have them sequence you in. (I should have thanked them but didn't; I dialed in Champaign tower and waited. After a few minutes, I initiated contact.)

Champaign tower, this is Cessna 80233 on a heading 060 degrees about three miles from your airport, but I don't have you in sight, over.

80233, Champaign tower, we have the report from approach; you are forty-three miles southwest; just keep on the heading and we'll contact you again.

Forty-three miles! How did I get that far off course? Well, nothing to do but fly and wait for Champaign to come into view. The tower came in a couple of times and told me how far I was. Finally I saw the airport and landed, totally safe. Actually, during the entire experience, I never really felt danger for myself or the aircraft; anger and confusion, yes, but not the kind of danger that would cause real panic. Yet I had broken down to the extent that, as Tom was to point out, I had neglected to use some pretty obvious and very available alternatives. I was far enough southeast by the time I had called for help to have tuned in Champaign on the working VOR radio and simply flown right to it. I could have done a better job of watching the ground landmarks and sensing when something was initially wrong and deduced what had gone amiss. I could even have landed at any one of a number of small airports and found out where I was; in effect, utilize the closest thing to stopping at the local gas station. But I didn't. I had become confused and disoriented and reached for help. Why, however, had I gotten myself into this? The explanation was to come as soon as the trip back to Midway was completed.

The trip back was uneventful except for experiencing some radio difficulty. I did contact the Chicago Flight Service and extend my ETA (Expected Time of Arrival) since I had used one hour being lost. Once back at Midway, I hoped Tom would be there so that we could discuss the flight and perhaps he could shed some light on what went wrong. He was there and the explanation came quickly. "When you dialed in the Victor Airway radial of 115 degrees, are you sure you didn't put in 150 degrees instead?" It hit me; that's it, that's what I must have done. But why couldn't I find my location visually? "Because you probably flew off the map!" Before I left, Tom had asked whether I had the St. Louis sectional map, reminding me that my route was close to the bottom edge of the Chicago sectional and saying that Rick once had a student who flew off the map and didn't have the adjoining one. Now Tom had such a student and I had the other map but didn't use it!

By tracing the 150 degree radial from the Peoria VOR, it became evident that I had flown far southwest, placing me precisely where I was found, some forty-odd miles southwest of Champaign. The airport I had seen was Decatur and was on the St.

Louis sectional; had I looked I would have recognized it from its size and shape as portrayed on the map. Tom then took me through all the ways I could have gotten out of trouble that had not occurred to me. He was obviously (and with good reason) upset with my performance. By the time our review was over, he softened with "Well, you made it, and ultimately you did the right thing—you confessed your need, complied with instructions, and got out of trouble. Now, let's talk about preparation for the license. That extra hour in flight, though you were lost, put you over the ten-hour solo cross-country requirement. Without it, you would have had to do another short one and undoubtedly what you learned in that hour was more worthwhile than just another trip." He was right. I had learned a great deal, but how had I confused 115 degrees with 150 degrees? This still troubled me. The trip, by the way cost $103.00; I paid extra to get lost!

9

The "Ticket" to Learn

Undaunted by having gotten lost on the long cross-country, I plunged ahead with the final hours of preparation for the FAA Flight Test. It occurred to neither Tom nor me, nor apparently anyone at the Flight Training Center (the "news" got around) that I would quit. Soon I heard other stories of getting lost or into one sort of trouble or another come out of the woodwork, apparently "released" by the comfort of sharing with someone else who goofed.

Having completed all of the stated FAA requirements—forty hours of flight experience including the solo and solo cross-country work (see digest for breakdown of required hours) the need was to sharpen flying skills as well as brush-up on aspects of ground school knowledge. Taking the later first, part of the flight test is an oral examination covering everything from the theory of flight to emergency in-flight situations; special emphasis is placed on knowing a great deal about your airplane, working a weight-and-balance problem, knowing its flight statistics and otherwise understanding its characteristics. About a week before the test you are given the destination for purposes of planning a cross-country flight, the idea is to plan your pre-flight work to that destination, with the final calculations made on the day of your test and gone over by the examiner. This complete "oral exam" phase is then followed by the flight test.

Once again it must impress the reader, as it did me, with the lack of similarity to the preparation and testing process for obtaining a driver's license. The piloting standards today are very high, the preparation intense, reflecting, as they should, the difference in skills as well as the risks inherent in even the casual operation of an airplane over that of a car. How then can anyone say, with any seriousness or authority, "If you can drive, you can fly?"

The flight test is detailed by the U.S. Department of Transportation, Federal Aviation Administration, in its *Flight Test Guide, Private Pilot—Airplane*, Revised, 1975; this booklet is available from government book stores, the U.S. Printing Office or any number of aviation supply sources.

With spring in full bloom, so were the droves of people coming in for their flight training; Tom and all the other instructors were booked ahead for two and three or more weeks. Tom had, literally, no time for me for another ten days and then only at 5 p.m. With daylight saving time in effect in Illinois, it was still light for a couple of hours past 5 p.m. Not wanting to lose the momentum of training, I found a couple of hours available with Rick, the Chief Pilot and scheduled myself in with him.

What a workout he gave me! Slow flight, approach-to-landing stalls, takeoff or departure stalls, turns about a point, S turns across a road and much more. I hadn't done some of these maneuvers since my first weeks of training. In the turns about a point, you are given a road junction and then you must fly around that junction in a circle, maintaining a constant radius by varying the banking of the plane to compensate for wind while maintaining a constant altitude. In the S-turns exercise, you must do a series of S turns so as to create equal half-circles on each side of the road, wings level when crossing the road, again without losing altitude. The obvious intent is to demonstrate your ability to control that aircraft with precision.

Then followed "hood work," flying the plane while a hood masked a view of everything but the instrument panel. As explained earlier, this training is essential for a would-be instrument-rated pilot, but some familiarization is deemed appropriate for all VFR pilots in the event you get into clouds or a weather situation where reference to the ground and horizon is lost. Try flying without these references—you immediately learn how dependent you are on the outside-of-cockpit view in your flying; this is not easy, let me (and everyone else who undergoes this training) tell you. And frightening, too. The plane seems to suddenly have a mind of its own, climbing, banking, accelerating, seemingly without your doing anything. But, of course, what you learn is that it *is* your actions and inactions that are the cause of the flight-path changes. The skill that must be acquired is to use the instruments, to trust them in terms of what they tell you and "fly the gauges"—purposely disregarding your "instincts" (which will be *wrong*) and relying on what the instruments state about the aircraft.

My hood work was not as good as it should have been; it seemed that the moment I got the plane in level flight, it seemed suddenly to be diving or climbing; stop the dive, and the heading was off; get the right heading, and the wings were no longer level. It was all too easy to fall apart, to become confused and lose control. Rick stated that most VFR pilots can't control a plane by instrument flying for more than ten minutes. I think he was generous.

"You're getting close; a few more practice sessions and you'll be ready," was Rick's assessment, adding that the FAA-designated test examiner used by the Flight Center had recently broken his arm and they were apparently reluctant about using someone else. Rick explained that while this man would certainly flunk anyone who didn't meet the levels of necessary proficiency, they were used to the testing techniques and methodology of this man; besides, it would appear that I wouldn't be ready until he was once more able to give flight tests.

In our next lesson Tom concentrated on hood work and gave me some valuable pointers on maintaining control; if you are told to climb and change heading, set the plane into a climb first he advised, then move the plane to that new heading slowly, very slowly as you climb, anticipating the new altitude by lowering the nose before reaching that point and doing the same on the heading, that is, rolling out of the bank before reaching the new direction. When I next flew with Rick and he had me under the hood, his comment was, "You've been practicing this, haven't you, and it shows." Indeed, practice and more practice is the name of the game.

It was in practicing a takeoff or departure stall during the next session with Rick that it happened—the plane veered sharply off to the left instead of stalling straight down. I recovered and thought I recognized what had happened. Rick asked, "What do you think that was all about?" "Seems to me the plane was starting a spin." "Right," he responded, "you were in a position for commencing a spin. How would you like to try some spins?"

I had read that at one time, a few years back, spins were required as part of the training for your Private Pilot's Certificate but were thereafter dropped. Most of the writing indicated that the dangers from practicing spins was greater than the value of the experience gained. Rick, however, believed that it was because too many people stopped their training rather than spin in the plane and that the light-plane manufacturers had successfully lobbied for its removal as a requirement. As we discussed the matter, I told Rick that I had frequently spun certain of my radio-controlled models, especially my P-47 Thunderbolt; it spun like a top. "How do you put it into a spin?" asked Rick. I described the nose up, power off, crossed rudder and aileron technique I used. "That's not quite the way we do it in a 152; take her up to 4500 feet, clear the path below and around, and we'll spin," he responded.

It took a while to get up that high, Rick pointing out that we would lose about 1,000 feet as we fell earthward. Reaching that height, we look carefully, clearing to all sides. Then the plane was put into a stall power-off position, the left rudder hit and the plane fell into a spiral turn earthward. Then, after another spiral, the twisting tightened into a spin. What a sensation! The plane spinning around what seemed to be a single point as the earth whirled around wildly. "Release the wheel; kick and hold opposite rudder," yelled Rick. I did and the plane promptly recovered. With power back on, we climbed back to 4500 feet to try it again, this time in the opposite direction. Once again, the feeling of facedown, earth wildly revolving as the plane accelerated in the spin, then recovery as the whirling stops short almost instantaneously.

The spinning of a model was always a "crowd pleaser"; gasps were heard when the model plunged, spinning toward the earth then suddenly snapped out of the fall and zoomed off straight and level. I don't know whether anyone saw us practice the spins, but as a "crowd-of-one," it was sheer excitement and thrilling to be in command of a spinning plane in the air instead of on the ground.

The training picked up in intensity. For some three weeks prior to the projected test date, I got out for dual instruction with Rick or Tom as often as I could. It was at one of these sessions that Tom asked whether I had scheduled any night flying. The FAA requires three hours of dual night flying if you are to have an unrestricted license; the failure to log those hours will result in a license that restricts you from flying after sunset or before sunrise. Knowing of this requirement and feeling that the challenge of daylight flying was plenty, I had made up my mind to accept that restriction and take the matter up at some later time. But Tom wouldn't hear of it.

"Three hours and you have an unrestricted license," he exclaimed. "Look, there will be times when you will want to get off early or find yourself coming home late and not want to put 'er down just because it's getting to sunset." Without further discussion, a time for night flying was set. Luck was with me this time; the night selected was perfect; calm, starlit, cloudless, and mild. In the interim, I had beefed up by reading about night flying; how emergencies were obviously so much more difficult to handle (try knowing where to land off-airport); how your health and eyesight were more crucial; how most pilots flare out for landing too high above the runway; how easy it is to fly into a cloud because they are harder to see at night. On the other hand, there is less traffic and in some respects it is easier to navigate since landmarks, towns, and roads stand out in such bold relief.

We had to wait until it was fairly dark on that beautiful May evening since it had to be logged as night flying. The only different instrument or tool, if you would call it that, was a small aviator's flashlight, one specifically designed for night cockpit work. Everything in terms of procedure was the same, except that light became all-important; checking the lights on the plane, in the cockpit (it's red) and distinguishing the different colored lights used on the airport runways, taxiways, and the like. But the procedures were all the same; preflighting the plane, communicating with ground control, the tower, all the same. Yet it seemed as though someone ought to say, "Lookout; it's night. Be careful 'cause it's dark out!"

Of course, no one did; this flying business is adult stuff where the pilot is responsible for all decisions; including the ultimate decision of whether to fly. The adventure began as I lifted the nose off the runway, pointing into the darkened sky. You must watch your gauges, the flight instruments, much more carefully because references to the ground and horizon are different and less distinct. The takeoff was smooth and we headed out west from Midway. Then it hit me, the strange beauty of the city at night, so different from the vista given from a tall building because it moved and changed, and so distinct from flying at night in a commercial airliner because your vision is so much more expansive and realistic than what you can sense through the small porthole of a 727. Most importantly, it is you, the pilot that is controlling that ever-changing view. You don't have to be a romantic—only a pilot (who perhaps are all romantics) to appreciate the special sensations that play across the mind as you gaze at the night landscape below you. Off to my right rear was downtown Chicago, with its tall buildings, lit up and towering like huge candles; lights flashing and the famous Palmolive Building (now the Playboy Towers) with its rotating floodlight beaming over the city. What a spectacle!

Down to work now as we plunged westward away from all the lights and signs of a city alive at night and into dark stretches with only occasional towns here and there; large highways so clearly discernible, but now more infrequent as they spread out from their urban confluence. We were heading out to Aurora, Illinois, and the task was to find the airport by pilotage, that is, by reference to the ground and map. I found a road that went reasonably close to the airport and then started looking for the rotating green and white beacon, the major visual sign of an airport. Once having spotted it (no small triumph), I called the tower and got no response; called again, still nothing. Then another pilot, having by chance heard the call, indicated that the tower at Aurora Airport had just closed and that they were last using Runway 27. Tom looked at a legend on the map and sure enough, clearly stated tower operating hours were listed and indicated it had closed down by this time. So, you're on your own in entering the pattern and doing your landing. Coming close up to the airport, I got a look at the runways and, sensing the pattern I would have to perform, began my descent and prepared to land.

Consciously I fought not to come in too high and it worked; my first landing was great. What an interesting sensation to come down through the darkness and put the plane down between the runway lights; it was a different and, to my surprise, pleasant experience. Up again and around the pattern we went, over and over again, but never the same as Tom changed the kind of landing he wanted; short field, soft field, no flaps. Then he got tricky and switched off the landing light, the light comparable to your car headlights and had me land by the runway lights alone; then he killed the cockpit instrument lights, but left the landing light on, then did a "finale" by having me land with no instrument lights and no landing light, in effect, demanding that I sense the height and speed of the plane as well as the distance to the proper touchdown point on the runway, all of which I did quite well. It was already apparent that both he and I were pleasantly aware that I was doing well, able to handle the night flying "curves" he was throwing.

"Enough here; take me to the DuPage County Airport," and we were off to the DuPage VOR (radio navigation aid) and from there on an outbound radial to the airport. The tower at DuPage was not only operating, but, as I was to find, extremely busy. It seemed that everyone was out flying that night; indeed, it was such a good night that undoubtedly a good many students and their instructors were getting that night flying requirement fulfilled. Now the challenge was to line up with other planes in the pattern and pay close attention to the controller's directions; he had so many planes taking off and landing as well as doing touch-and-go's that two runways were in use. I did touch-and-go's as well as full stop landings here, the latter to get practice at taxiing around an airport at night as well as taking off into total darkness, as against the more lighted environment of Midway, surrounded as it is by the city.

As the time drew to a close, we headed back southeast to Midway, but the training was not over. On went the hood and I had to fly for some time by instruments only. Then, with the hood off, find Midway which Tom indicated was "somewhere" around. A few turns and a lot of peering around and I spotted the large dark patch with the rotating beacon (closer up, I would see the criss-crossed runway lights) and came in for a landing. It had, in all, been quite an experience; from the training standpoint excellent—I had performed well. But more importantly, it was just plain exciting, a new and quite thrilling experience to pilot at night, to feel that special ambience of night, of flight, and of adventure.

The date was now set; three students would take the flight test on June 14. The usual FAA-designated examiner had sufficiently recovered from his broken arm to take on the three of us. What was he like, I wondered? An ex-Navy pilot and instructor I heard. Meanwhile, I got in those extra hours of preparation with Tom and Rick, whichever was available. Over and over again practicing the maneuvers called for in the test, only some of which (but which ones?) would be tested. One thing I did learn was that Wally, the examiner, was an absolute stickler for safety. So the training went: lift that wing to see what is under it, lower the nose as you climb to look ahead, do clearing turns (90 degrees to the right and then the left) before you start any maneuvers. Now, turns around the point—hold that altitude. Give me a departure stall to the left; you've lost your engine—what are you going to do; land on the numbers (the runway designation numerals); watch your airspeed—60 is not 65. Grind and grind again, but I both appreciated and actually enjoyed the challenge of sharpening my skills.

Tom suggested I go up by myself and practice, since I hadn't gone solo for some time, so I carved out a two-hour slot and flew off southwest. By that time I had been informed that my planned flight requirement would be a cross-country to Peoria, Illinois, a trip of about 110 nautical miles (about 130 statute miles); thus I practiced by going in the direction of that flight, knowing that I would not be completing the flight to Peoria but would be starting in that direction. I wanted to be familiar with the area. Flying over to a small airport that might be the one to which the examiner would divert the flight, I got into the landing pattern, doing touch-and-go's as well as full-stop landings to get the feel of the place. Satisfied that I was comfortable with the airport, I headed back to Midway, coming up to a reporting point where I called Midway Tower. No response. I called again. No response. What's wrong?

I checked the frequency I was transmitting on—correct and cranked up the volume. This time the tower said, "I hear someone out there but the signal is very weak; can't make you out." Once again I called and identified myself—no response. Tried again; by this time I was close to the airport and unless I could communicate, I would have to get out and away since you can't land at a controlled airport without being in communication. Though that communication could be in the form of signals from a light gun (this rarely used procedure is available where you indicate by your flight pattern that your radio is dead and the tower sees you). In my case, I could hear the tower, but he could not hear much of me. Then the controller said, "I can see you Victor Lima, (he had caught the letters on the end of my airplane identification number) but I can't hear you. Clear to land, 22 Right." He had seen me, recognized my problem and immediately cleared me to land, keeping other traffic out of the way.

Once on the ground, the tower was able to hear me and said, "Better get that radio fixed," I'll say, I thought and thanked him. I reported the trouble back at the Flight Center and they took the plane out of service. I later learned that the trouble was the microphone, which was immediately replaced. Had the tower not seen me, I would have had to go to an uncontrolled airport and call for someone to come and get me since I couldn't return to Midway with that plane.

At the end of our last dual session, Tom said, "Fly as well as you did today and you'll get that license." With that confidence building evaluation in mind, I plunged into book studies for the oral examination portion of the flight test. Hours were spent reviewing all that I had learned in ground school training, with special emphasis on knowing that Cessna 152, the airplane I would fly for the test. I had to know all of its characteristics such as weight, speeds, fuel system, and flying limitations, all of which could be the subject of questions on the oral phase of the flight test. The trip to Peoria was planned with special care and detail. The administrative forms necessary for the flight test were completed and then reviewed by Tom. The engine and airframe logs which are kept on every airplane were reviewed as well as the documents of ownership on the plane. Then, with student license and medical certificate in hand along with the test results from the written examination and the hours in the flight log counted, all seemed in readiness. But would the weather cooperate?

I scanned the weather reports with eagerness and antici-pation, if not anxiety. What a blow if it had to be rescheduled with the resultant repetition of the psychological buildup. Every indication was that June 14 would be a good day, and it was. I arrived at the Flight Center an hour and a half early, just to go over everything, especially the documentation, once again. I got the weather briefing for the trip to Peoria (that I knew I would not complete) and with the wind-aloft readings, made the final necessary calculations including the use of Wally's weight for the weight-and-balance problem. Any airplane must be within strict weight limitations (not overloaded) and that weight prop-erly balanced for safe flight within what is known as an "envelope" of weight distribution. Preparing the calculations necessary to determine whether we were overweight or unbal-anced includes the incorporation of the pilot's and passenger's weight. Wally's weight was reported at 210 pounts, though, it was quipped, that was probably somewhat under-estimated.

Tom had re-arranged his flying schedule to meet with me and make certain all was in order; this was a courtesy and thought-fulness typical of the man and the Flight Center; it was deeply appreciated. We went over my log book and sat discussing what I might expect. Then it was time for me to go and since Tom wanted to be certain the information on the exact weight of the plane was known from documents kept in the aircraft, he drove over with me and made certain all was in order.

In a small room set aside for the purpose at one of the hangers at Midway sat Wally. Mid-fiftyish looking, stocky but well built and with close-cropped grey hair, he certainly looked the role of the ex-Navy pilot instructor. Favoring his mending arm, after a brief hello, he plunged immediately into review of the documents I presented. Two things were readily apparent: he was a very old hand at this flight-test business, and he would waste not a moment in moving from one point to another in the examination process.

After correcting the fact that I didn't spell out my middle name but used only my initial, he reviewed my logged hours. A scene flashed into my mind—what must his log books look like, if indeed he had bothered to log all those thousands of flight hours in different aircraft? How pitiful must my few pages appear! Wally's staccato voice pumping questions brought me back with a jerk. Within a matter of a few minutes, he covered a range of subjects, each time moving to a different area the moment he sensed my proficiency as acceptable. He corrected me on only one point where I had mentally subtracted instead of adding. Without ceremony and within what seemed to be a matter of minutes, he rose abruptly and with a "let's go," we went out to the aircraft.

I was somewhat confused; that's all to the oral? Why he hadn't touched very deeply anywhere and the whole process was so short. But no time for such evaluation thinking now; now it was time to do it right in that aircraft. Remember, I told myself, take your time, check everything. Remember to start your stopwatch at takeoff; remember to adjust the ailerons as you taxi. Now for the pre-flight, I murmured to myself as we approached the plane, show him you know what you're doing.

But almost none of these pre-conceived ideas of how I would respond to the flight test were to come about; Wally had his own ideas and was used to giving orders, which were to be promptly obeyed. I said I would taxi over and fuel up (that's showing the proper caution); Wally said there is plenty of fuel (there was) so let's go. He also added that the plane was fine since it had just been pre-flighted and flown. Was he telling me not to bother to pre-flight? Aha, a trap of course, I said to myself—this safety-minded guy wants to see whether I'll skip the process. No way; I'll show him. So I went through a most careful pre-flight (which is the right way in any event) including climbing up to look into the fuel tanks to indicate I knew that a cautious pilot doesn't trust the fuel gauges. No sir, I took my time, certain that is what he really wanted. In retrospect, I'm not so certain.

Having completed the pre-flight, I climbed in and went through the start-up procedures, carefully going through the steps, but not careful enough because I forgot to turn on the rotating beacon light that sits on the top of the tail; it isn't listed in the daytime procedures but had been taught to me by Tom as an added safety measure; Wally simply leaned over and snapped it on. I started taxiing to the active runway (the one for takeoffs and landings) which happened to be on the far opposite end of this large field. The next shock was when Wally told me to taxi quickly since it was so far away! What about all the precautionary training about slow taxiing? Again, did he mean it? This time I thought he did, sensing his impatience and I taxied faster by quite a bit than I normally would.

"Give me a short field takeoff," commanded Wally as I ran through the checklist prior to takeoff, having reached Runway 22 Right. "Yes sir," I responded, moving the flap handle to 10 degrees and mentally running through my mind what I had to do—get the nose off at 50 knots, climb out at 55 knots. The procedure is designed to get the plane off the ground in the shortest possible distance, the idea being that the field is short and you have to get the best angle of climb out to safely get off the ground. I handled it well, I thought; Wally said nothing.

With the plane nosing into the air and banked gently onto the heading that I had calculated was correct for the trip to Peoria, Wally instructed me to "climb to 2,800 feet and give me a ground speed check in ten minutes." A ground speed check requires timing and I had forgotten to start my stopwatch; I clicked it in and calculated what I would have to add to that time. I lowered the nose every few hundred feet to see ahead and kept my "head on a swivel," the expression used to describe the need to search the skies constantly for other aircraft in order to avoid traffic. I also wanted to demonstrate to Wally my safety consciousness. Then as we were continuing to climb, he told me to lower the nose once in a while to see ahead; lower the nose? Hadn't he seen that I had been doing just that? I responded that I had been, but made the movement even more definite.

When he asked for that ground speed check, I reached for the flight calculator, but Wally insisted on doing it with me by mental calculation, attempting apparently to show me it could be easily done without the calculator. I answered as best I could giving an estimated ground speed which is almost invariably different from what the airspeed indicator instrument reads since the latter reflects speed through the air in knots. Then Wally told me to change course; we had just passed the first checkpoint, and apparently that was as far as we would travel to Peoria.

The meaning of this command was clear; it signaled the end of the cross-country phase and the commencement of that part of the flight test that would examine my ability to control and maneuver the aircraft up to the level of performance demanded by the FAA-promulgated standards. The next forty-five minutes moved so quickly and contained so much that it is practically impossible to delineate all that took place. When it was over and I was sitting with Tom back at the Flight Center, he asked me to brief him on all the maneuvers and other demands I faced during the flight phase of the test. I began by recalling that Wally called for minimum controllable airspeed at a certain heading, keeping the existing altitude. After that, I recalled a number of maneuvers and then would pause in my recital. "That's all?" Tom asked. A moment's thought would bring forth more, then another pause as I thought I had recited all. This went on a number of times; after just a little prodding I would come up with several more items. Then Tom would ask, didn't he have you do this or that and almost invariably, I responded that, oh yes, I forgot. What I came to realize was that in those minutes following the turn-off from the course to Peoria, Wally had, in rapid-fire succession, put me through an incredible number of exercises, never pausing for more than a moment (and sometimes without any pause) before demanding another. Once again, his technique in the air was apparently the same as on the ground; if he saw that I could handle the exercise, perform the maneuver or react properly to the demand he made, he terminated immediately or abruptly aborted what I was doing.

Three kinds of stalls, turns about a point, responding to a loss of engine operation, turning the plane while at minimum controllable airspeed, maintaining maximum glide speed while descending, these and many more were demanded in rapid succession. Then came radio navigation work, flying to and from VORs (radio navigation transmitting stations) followed by what seemed a long time under the hood. All the while, Wally said not a word of comment, good or bad. All he did was apply pressure, bark commands and demand prompt, effective, and correct responses.

Suddenly, he slowed down. With just a slightly different tone, he told me to take up a northeasterly heading and maintain the assigned altitude. I recognized that we were on the way back to Midway by the heading he had given. Suddenly, Wally seemed to take over, calling the tower, reporting our location and telling me what to do. Just as we approached the airport, the tower called us and said there was a wind shift and they were switching runways. I banked the plane around and Wally told me to do a short-field landing, meaning one which you touch down as close to the beginning of the runwy as possible. Yet, though he was telling me to do it, he seemed to resist my performing the steps I felt were necessary. He wanted me to land close enough to turn off on the first taxiway, but I didn't make it; I had to turn off on the second.

Moments later we were parking the plane, and without a word Wally deplaned and disappeared into the hangar area that contained the office where we had begun. I went in and found the room empty. After a while, feeling perhaps I was in the wrong area, that he had assumed I would be elsewhere, I went looking for him, only to find that he was sipping some coffee and talking with some hanger personnel. With a "be with you in a moment," he waived me back to the room and I returned and waited. Soon he appeared and again, without a word as to my performance other than saying my short-field landing could have been better, adding however that he thought that there was something of a tailwind, he took out his old battered typewriter and typed out a form, which as I correctly assumed, was my Temporary Airman Certificate. After reviewing it, he asked me to sign it and made the necessary notation in my Pilot's Log Book. Then I offered my check, which he stated should be in the amount of $35.00, shook my hand and asked if I would send the next fellow over.

As I journeyed back to the Flight Center, I felt more confused than elated. That was it? Just over one hour? After all of the intensive training of the past few weeks, not to mention all the months of studies and flying? As I mulled it over, a different perspective came to mind. It was obvious that Wally put a lot of trust in the crew and the level of instruction at the Flight Center and with good cause. It isn't likely that an instructor will allow a student to take the test until that person is really ready; what's more, the instructor certifies that the student is ready and competent, thus reflecting, should there be a failure, on the instructor as well, perhaps more than on the student. If you fail, the examiner gives you a statement of what areas must be re-examined; you then retrain in those areas and after thirty days you can take the test again. As to the length of the test standing in such clear contrast to the intensity and length of legal training, wasn't it true that after fifteen exhaustive weeks of study in four or so law courses, the determination of pass-fail is on the basis of one short, relatively speaking, test? I would guess that had the examination been with an FAA examiner, that is, an employee of the FAA rather than a testing designee such as Wally, the procedure would have been much elongated and "by the book"; it would, by the way, be free since there is no charge for the flight test taken with the FAA examiner.

The scene at the Flight Center was much better; handshakes, congratulations and pride were all abundant. During the debriefing earlier described with Tom, I spoke of my disappointment with the flight test. He said he understood and felt it was reacting quite normally to the intensity of preparation and the desire to show the examiner that you really can do it. Then it was time for me to go. I thanked Tom and Rick for their help and scheduled myself with Tom for the following week for a checkout flight in a Cessna 172, the Skyhawk. It had been over ninety days since I had been checked out in that aircraft, so I wanted that flight to make certain I had retained my proficiency.

Eleven months, ninety-five hours of flight time, two hundred hours of study that included listening to tapes, reading texts, and doing lessons had elapsed since I first walked into the Flight Center. As I walked out with that coveted license in hand, I could not but reflect on the dual responsibility it carried; I was now found proficient to pilot an airplane and in doing so have other people, as passengers, rely in crucial fashion on my skill. Just as relevant was that I had only earned a ticket to learn, carrying with it the obligation to sharpen my piloting skills and to become an ever more competent aviator. I looked at the certificate again and what it told me was the joy of flying is earned through fulfilling the obligation to be really good at it—nothing less than a pro.

10

Should You Fly – Can You Fly?

Statistically, it shouldn't be; the improbabilities are too great. Out of a faculty of fewer than forty-five, this law school has five pilots! All of whom, except me, had been flying for many years; most had advanced ratings (instrument, multi-engine, etcetera) of different sorts. Thus, occasional "hangar" flying was a definite probability in the halls, offices, and lounges of our institution. One pilot member had organized an Aviation Law Society, devoted to interesting students in the study of aviation law and bringing leading authorities in the field to meet with the group.

My next-door neighbor, a younger and newer faculty member had learned of my having undertaken flight training, occasionally stopping to chat about my progress. After reading a portion of the manuscript of this book, he came in to talk over my work. His basic response was that, "You certainly have captured the feeling of what it's like to learn to fly. I had really forgotten so much of what you relate. It was such a long time ago for me." He had learned to fly when he was eighteen and was now in his early thirties, an instrument-rated pilot with a good amount of experience. Quiet, soft-spoken, intelligent, he exuded an understated air of competence and integrity.

The conversation turned to safety factors in flying, I stressing that my desire was to give a balanced, realistic view to the would-be pilot. My colleague endorsed the idea and then added, "You know, I learned to fly working with four flight instructors, three of whom are dead, all three killed while flying." I was stunned and sat silent as he continued. "And the amazing part is that each was the kind of pilot you would have no hesitation about flying with—all had tremendous flying backgrounds, loads of experience." He then proceeded to rattle off, in brief, the impressive flying background of each and a brief synopsis of the death in flying of all three. One in particular stuck in my mind because it was close in terms of location (a nearby airport) and had the element of the totally unexplained. He was coming in for a normal landing with a student; it was a bright, clear day, and on final approach to the runway, the plane "seemed to fall out of the sky, forty feet above ground, nose first and killed him."

How did he handle the impact of the deaths of these instructors? He continued to fly, but most often alone. He would only rarely fly with his brother (who was also a pilot) nor with anyone, if he could at all avoid it, who was a father or mother of young children. He was apparently willing to risk his own life, but not that of those whose death would bring excessive grief and tragedy to the living.

Certainly the story is exceptional, the odds involved in the deaths of three of four of his instructors was quite freakish. Yet it seems to accentuate a valid point for anyone considering learning to fly; the safety-fatality risks must be considered, must be faced up to if you will, in making an intelligent and knowledgeable decision about private piloting. Just what are the facts about the safety-fatality aspect of flying your own private plane? Some thoughts have been delineated in chapter 4. Here the matter will be nailed down in terms of the latest facts and figures available.

The National Safety Council did some studies on the comparative safety of different modes of transportation using government (including FAA data) sources. The comparative death rates per one hundred million passenger miles derived from material for the years 1970 through 1974 were as follows:

> For scheduled domestic air-transport planes—0.10
> For passenger cars and taxis—1.82
> For personal aviation—48.47

There were 4,000 fatalities in personal aviation during this period which they calculate produces the figure of 48.47

fatalities per hunderd million passenger miles.

More statistics: In 1977 there were 693 fatal general aviation accidents (non-commercial flying) down two from 1976. The total number of fatal and non-fatal general aviation accidents in 1977 was up 6.7 percent to 4,476 with the overall accident rate increasing slightly in 1977 over 1976. These figures are taken from the National Transportation Safety Board.

The accident record through mid-1978 grew even worse. A 20 percent increase in general aviation accidents has been reported, compared with the same period in 1977. The concern is to be found in the fact that, while the number of flying hours have increased and thus the death rate is to be anticipated as increasing, improvement in flight safety comes only when the number of accidents remains the same while the number of hours and miles flown increases. The evidence is clear that this is not happening, at least through mid-1978.

The Federal Aviation Administration has found that the causes of accidents are about the same, just more of them taking place. Ironically, midair collisions did not increase in proportion to other accidents reported during the first five months of 1978 in comparison with 1977, in fact decreasing from 14 to 9. It's ironic because the worst midair collision took place in October of 1978 between a 727 and a Cessna 172 near San Diego.

In court, in a recent accident case tried with respect to a Cessna aircraft, statistical testimony was produced and yielded these figures in terms of fatalities per passenger mile:

> General aviation or light-aircraft travel is approximately eight times more dangerous than private automobile travel.
> Private automobile travel is approximately ten times more dangerous than commercial travel.

Further testimony revealed that approximately 5 percent of all Cessna aircraft (the leading manufacturer in terms of number produced) in any given year would be involved in either a crash or reportable accident. This percentage of accidents translates into this statistic: over the twenty year life span of a Cessna aircraft, it has a 60 to 70 percent chance of being involved in the sort of accident that could produce serious injury or death.

Pilot error, weather, mechanical failure, midair collision, lack of proper restraint systems (seat and shoulder belts) design defects, drinking while flying—these and many more "causes" could be tabulated and analyzed, but there are two cardinal messages to the would-be pilot:

1. Don't fool yourself or anyone else (your family, potential passengers) about the reality of danger in flying. It's there and must be taken into account in your decision to learn to fly. It *is* more dangerous than driving or riding as a passenger in a car.

2. If you decide to learn to fly, make up your mind to become really proficient and competent as a pilot—nothing less. Accept the fact that you can't just do a half-way job; that you must truly dedicate yourself to consistent study, practice, and increased proficiency. You can't "closet" your license like you can golf clubs and tennis rackets and expect no more than some strained muscles when you "uncloset." In learning to fly and in flying after you get your license, you must consistently aim at becoming more proficient—your life depends upon it.

That these attitudes are truly demanding ones, too demanding for most, is revealed by the fact that only about 40 percent of those who take out student pilot licenses ever finish their pilot training and get their Private Pilot Certificate. The dropout rate is even higher when you consider that most students don't go for their physical and get their Student Pilot Certificate until they are fairly well into flight training. What's more, the realities of these demands come through in the figures that state that about one-half of those who get their Private Pilot Certificate quit flying within 100 hours after achieving their goal; this dropout rate is particularly revealing. There are about 750,000 licensed private pilots in the country, a figure that has remained constant for several years. Under the GAMA promotional program, approximately 10,000 students start per month, yet the dropout rate among students and among licensed pilots is so great that those new pilots entering aviation barely balance, it is reported, those who quit.

One reason apparently so many licensed pilots quit early on after becoming licensed has been alluded to throughout this book; the severe limitations put on the VFR pilot by the weather. Plainly put, you can only fly when the weather is really good, not marginal, and when that good flying weather also exists where you intend to fly. For most of our country, weather instability and change are the rule, not the exception. Therefore, patience, delays, disappointments, and frustration plague the VFR flying scene. The way out, the ticket to expanded flying capability is through the instrument rating. When you have that instrument rating, unless there is weather so severe that you can't take off, go around it, or land at your destination, marginal or poor weather conditions don't stop you. For example (and this is not at all unusual), the point of departure may be below VFR minimums (1,000 foot ceiling; 3 miles visibility) but up above, say 6,000 feet, it's clear skies. The VFR pilot goes through the low-lying clouds near the airport and up into those clear skies. The problem is that FAA instrument rating requirements are high (two hundred hours, hard written exam, plus flight test) costly, and difficult. Not all private pilots, even quite competent ones, can master the special skills involved. So, with only 30 percent of pilots instrument rated, most pilots have severe limitations on their flying. With these facts in mind, another justifiable statement to the would-be pilot is as follows:

3. The limitations imposed by weather on the average licensed private pilot are such that flying can only be done in really good weather, imposing restrictions on the amount of flying that can be "counted on" in terms of a definite time or the ability to go to a definite place.

Turning to another aspect of flying, it should be stated that anyone undertaking flight training should have demonstrated to themselves and their instructor that they have the physical coordination, general physical attributes, and psychological makeup necessary to achieve competency as a pilot. Man is an earthbound creature and not particularly suited for any environment other than a life managed with feet or wheels on the ground; essentially a two-dimensional existence. Consider this: for untold centuries man has moved onto the rivers and seas in all sorts of crafts for serious as well as purely pleasureable reasons. Those who have devoted themselves to this form of transportation, the sailors of the world, have mastered the skills required, learning from countless generations of those who have preceded them. Two comparative aspects are involved. Travel by water is still essentially travel in a two-dimensional world (perhaps something of an exception being the movement of a submarine). Secondly, unlike travel over water, air travel in heavier-than-air ships is only some seventy-five years old, an infant, barely alive in comparison with travel by water. Obviously, generations of air-travel experience will supply the knowledge and experience concerning flight that is lacking today.

What is not obvious, but is incredible, is how quickly man has conquered this new, three-dimensional form of travel, especially when one considers that the best Wright brothers flight in 1903 covered 852 feet and lasted 59 seconds. Today, well, records for speed and endurance continue to be broken and boggle the mind, not the least of which is man's ability to leave the hold of the earth's gravity, to engage in space travel. The average person in this nation can speed across the land at 600 miles per hour in a commercial air carrier, enjoying the advantages of speed, comfort, and safety. But it is still the newest frontier, this business of flying and for those who would taste of its pleasures, thrills, and conveniences, there is the realistic need to amply answer the physical and psychological demands of this least hospitable, least natural environment for man, the air.

Piper Tomahawk

The Tomahawk's Lycoming engine (above) A Tomahawk (below)

Off on adventure

So it is not without reason that the would-be pilot should examine whether the physical and psychological demands that do exist, that must be met, are carried by one who would learn to fly. To suppose (and, as noted, it has unfortunately been often stated) that *anyone* can fly, that in effect *everyone* has the ability to learn to successfully (again that ultimately means safely) fly is not only wrong as a proposition, it is *dangerous*. Those who have learned to conquer movement in three-dimensions, to control that machine in an "unnatural" environment have indeed included what might appear at first blush to be an unlikely group: people with physical handicaps (one-eyed, lacking a limb) and certainly as well, people who were of marginal intelligence and intellectually generally ill-equipped. But that doesn't mean anyone can fly. Whatever deficiency or handicap such a person may have had, what they did have was the right combination of physical and mental traits needed to learn how to fly. The wide variety of people who have succeeded give witness to the idea that you can't tell from looks, sex, age, or occupation who will make a good pilot or not; only making a sincere effort to learn will tell. Therefore, to this list of basic statements for the would-be pilot is added:

4. Not everyone can (or should) learn to fly. It is a skill requiring the right combination of physical and psychological traits. Only through the test of a sincere effort to learn to fly will you know whether enough of those traits are there in you in the right combination.

The physical demands in terms of coordination, agility (consider the number of dials, knobs, and switches alone that must be moved), depth perception, and the like must be met. These demands are far in excess of what is demanded in driving a car. As might be expected, the psychological demands involved in flying, while less tangible, are just as crucial. The stresses are often intense; judgments, the right judgments, must be made, not just once in a while, but constantly and consistently. When to begin this procedure? At what point do you enter the pattern for landing? When do you cease that procedure? At what point do you recognize danger and how do you react? It is probably what can be lumped together as "judgmental capacity" that most cogently defines what is required of the pilot, keeping in mind that the test of your capacity takes place in that most unforgiving crucible of a cockpit, moving through the third dimension, space.

People will often ask pilots or pilots-in-training whether they are afraid. If the thought of you in an airplane, hundreds to thousands of feet in the "thin air" and moving at twice the speed you have driven a car frightens you, if you are definitely uneasy about the thought of a single engine and if you experience real fright during your first lesson or two, then flying is not for you. There is plenty to be frightened of; if the entire process of piloting a plane becomes nightmarish (as it has for many who try it) it only means that private piloting is not for you. For those who love to fly, they will most often admit of concern as distinguished from fear.

They are concerned—about the plane, the weather, their skill, their passengers—concerned about everything related to flying because they are intelligently and consciously cognizant of the dangers of flying and the need to fly will all factors in their favor—concerned but not afraid, because if you are afraid enough, you won't fly. Competent pilots are not daring, bold, or dashing and rarely wear helmets, goggles, and white silk scarves. They love flying and want to enjoy the ambience of flight, so they prepare and check and check again so as not to spoil "their thing" by blunder, incompetence, or some failure that could have been avoided. A frequently posted sign found around flight centers reads

> *There are many old pilots and*
> *There are many bold pilots,*
> *But there are no old, bold pilots.*

For those who want to learn to fly, sound advice is to get the best flight training available and affordable. The digest following this chapter summarizes how and where you can locate such flight training. Learning to fly is not like learning most other skills; you really have to take the matter seriously. That doesn't mean it can't be fun (in fact, it can and should be) but it does mean that the decision to learn should not be a snap decision, a "lark" if you will. You should check out the flight training center, ask questions about how many planes they have available for training, how many instructors are on their full-time staff, whether they have hangers for their planes, and then get to talk to a few student pilots for their evaluations. You should try a demonstration flight in a low-wing (one or more) as well as a high-wing plane to see which one you prefer.

Once you opt for flight training, get involved, be dedicated and learn all you can. Don't rush for minimums—in terms of hours of logged flying time or skimp on ground school studies. For the would-be pilot:

5. Learning to fly is important enough to choose flight instruction with great care. Because your life and the lives of your passengers will be at stake, your attitude and involvement as a student pilot must reflect your insistence on achieving the highest level of competency you can attain. This must be your first priority—not "how quickly" or "how much money."

This is not to cast any kind of pall over the endeavor; as my personal account reflects: thrills, pleasures, satisfactions in learning to fly as well as a conscious knowledge that your life has changed, that you have dimensions very few have because you are a pilot, that you experience that which *only* a pilot can experience—all of this is true and all awaits the student pilot. The emphasis is on the realities of learning to fly, which if considered, accepted, and met head-on will help insure the making of a rational decision as to whether flight training is for you, avoid disappointments in terms of unreal expectations and hopefully encourage the right attitudes about flying competency. The joys and personal fulfillment in becoming a pilot are there with a full knowledge of the realistic demands of flight training and the limitations of flying for the average VFR pilot. I saw it said in a few words printed on a coffee cup at my flight training center, "If you ain't a pilot, you're nobody." The point is that those who believe that also know what it takes to become and remain a good pilot.

A factor often overlooked about undertaking flight training is your need for support. This support takes on a number of aspects, one of which is that you need the encouragement of your parents, spouse, close friends, and even, in some cases, business associates. If your position or relationship is such that you may not have that support, the decision to learn may generate problems or confrontations with important people in your life that will make moving ahead through flight training difficult, painful, or, perhaps, impossible. Knowing that you will need this moral support, seek it, and if you don't get it, weigh the implications carefully. Perhaps you can encourage that person to read this book with the idea that he will not only get some insights and understanding about learning to fly, but also will realize that you have a sound idea of what is involved.

Another form of needed support is expressed in terms of money and time. You need both in amply supply. Elsewhere it has been stated that given the average number of flight hours and incidental costs plus the inflationary factor that currently prevails, it will cost between $2000 and $2500 to learn to fly. This figure should include ground school costs and equipment.

Some students buy a course involving a package price but that package has a definite hour limit, usually less than the average student will require. Watch out for "deals" which give any kind of set price; be certain you understand how far that money will take you in terms of both air time and ground school. Unless there is a genuine savings of substantial proportions (where that saving is clearly demonstrated to you) the packaged courses or special deals are not worthwhile, especially when you are just beginning flight training.

How do you know you will like the experience and want to complete your training? To those who would answer that they know they are committed, this has got to be the worst kind of justification for putting yourself through flight training. What happens to your money if, after say five hours, you decide to quit? What if sickness intervenes or you can't pass the physical exam? From all of these vantage points, you are better off to "pay as you fly." Often you can deposit a sum of money and charge your instruction against that deposit. Otherwise major bank and credit cards are accepted along with checks and cash.

Be certain you understand who pays for fuel (is your price per hour for the plane "wet" or "dry") and who pays the parking or landing fees. As of this writing, fuel at Midway is about $1.00 per gallon, so who pays makes a difference. On of the cross-country solo flights I have described, flying solo for 2.9 hours cost $60.90, with my school paying for fuel and the landing fee. Remember that the minimum is ten hours of solo cross-country flying; the dual cross-country is, of course, more expensive since you're paying for your instructor's time as well. In summary, you have to consider the financial commitment and budget for it. Under most circumstances, you are best off to pay as you fly.

Similar to the matter of money, the personal commitment of time has been explored in this book. In substance, twice weekly flying lessons are recommended of approximately two hours each. If you can only fly on weekends, the process is slower and more difficult because of this limitation, keeping in mind that the greatest competition for lesson time slots is on the weekends. Weather may also deny the opportunity to fly and you may face weeks of weekends of no flying. One way around this situation is to somehow get some time in during the week, perhaps before or immediately after work; during the longer summer days this may be possible. Another alternative is to take vacation time to learn to fly in a two or three week intensive training period at a resident school. The digest lists a number of such resident flight schools. Still another alternative is to convince your employer that your flight instruction will be of benefit in your work (and that is certainly true for an amazing number of businesses) and to allow you time off during the day for instruction, perhaps "charging" the time against your vacation at the rate of one-quarter of a day for a block of hours off.

You must also consider the fact that whenever you take that time, it's time away from your family, friends, and other activities. Once again you need their support, not hassle about the time devoted to flying. Furthermore, your ground school training, both at the flight center and at home is important and time consuming. You have to "hit the books" and really study. That study should also include reading magazings and government publications about flying. The two hundred hours of study is not at all unreal, and that can reign havoc with family, social, religious, and business activities.

For the would-be pilot, it can be stated:

6. To undertake flight training successfully, you must have the moral support and encouragement of those who will be affected by your flying and associated studies. You will have to be ready to expend the monies, to find the time for flight lessons on a consistent basis, and to devote many hours at home to study.

Should you learn to fly—can you fly? Your careful consideration of the topics raised in this chapter should help you decide; to help you nail down these considerations even further, take the self-test that appears as appendix 1. Then, with enough positive indications for you, personally, in terms of all that has been explained, use the guidance contained in the balance of the appendices to locate the right flight training for you.

After a few lessons, sit down with your instructor and ask for an honest appraisal of your potential as well as your progress. Then ask yourself if being a flight pilot is what it has meant to hundreds of thousands—a way of life and a part of their life that they could not not do without. If you are moved to go ahead, to finish your training, then demand nothing less than a real sense of professionalism in your flying. You owe that to yourself, to those who trust you enough to fly with you and to your fellow pilots. If, after reading this book, and perhaps getting some introductory lessons, you find you either don't have the right combination for achieving proficient piloting, or you simply decide not to continue, you will have at least made that decision after examing the issues involved and "looking in the door" of the world of the private pilot; that alone is a worthwhile experience.

Digest on Learning to Fly

Appendix I
Quizzing Yourself

Appendix II
Basic Steps

Appendix III
Where to Go to Learn to Fly

Appendix I
Quizzing Yourself

Anyone seriously considering undertaking flight instruction has to ask and answer relevant questions in order to reach an intelligent decision. The following are questions and explanations that should help you focus on this decision-making process; these are not the *only* questions; every individual must examine his or her own particular situation and get enough positive feedback to warrant going forward on learning to fly.

1. Have I got the time necessary to learn to fly?

The process of learning to fly should not be stretched out over too long a period of time, hopefully (depending on the seasons of the year and weather conditions) not more than six to eight months. You should also plan on two lessons a week which will call for roughly a time block of two hours for each lesson once you arrive at the airport. Your home studies and ground school work are vital during this training period and time consuming as well. In all, a sizeable dedication of time is indicated.

2. Can I afford the cost?

The costs involved in learning to fly are not inconsequential; the range is generally in the area of $2000 to $2500 for the average student. In most instances, the money can be paid on a lesson-by-lesson basis although some programs offer a "package deal" which does constitute the immediate outlay of considerable money. Furthermore, other costs not incorporated in that figure include a certain amount of equipment and books that you will want to buy along the way. It must also be understood that once you have your license, flying is *not cheap* as a hobby or particularly inexpensive as a means of travel. Most private pilots rent their planes on an hourly basis, which is not cheap, since the cost of buying and maintaining an aircraft is high. While a gallon of gas will take you further in an aircraft, as of this writing, major airports are charging up to a dollar or more per gallon for aviation gasoline.

3. Can I accept being a beginner?

Are you ready to pay the "emotional" price of being a beginner? Can you take criticism—perhaps severe—perhaps from someone younger than you? Can your ego handle mistakes and come back for more?

4. If I quit or am advised to quit can I handle that?

Learning to fly is generally not a "closet" subject; your friends and family will know and share the excitement. Will you want to handle the scene when you quit somewhere along the line, whether it's your decision (for any number of reasons) or because you're advised to do so?

5. Am I too old?

Age does count; other considerations aside, the twenty-five year old is going to learn a lot faster than the fifty year old. It will also be more costly for the older person because that person is likely to be slower in the learning process. In some respects, age also works against the safety factor because your reactions are slower and your ability to learn and absorb are not as good. On the other hand, maturity is a "plus" in some aspects of flying, particularly in terms of ultimate judgmental factors where the more mature person is less likely to take undue risks.

6. Am I afraid to fly (perhaps I should be)?

An honest appraisal should be made and this question faced up to. There are significant statistical dangers involved both in learning to fly and for the private pilot. While civil aviation death rates have declined steadily since 1945, preliminary figures from the National Transportation Safety Board and Federal Aviation Administration indicate that for the year 1976 there were 28 accidents resulting in 4 fatalities for commercial aircarriers in the United States as against 4,567 accidents resulting in 636 deaths in general aviation.

7. Am I physically able to learn to fly?

You must pass a physical examination and if you are on medication for some chronic disease, you may have a problem. Therefore it is wise to get that medical certificate as quickly as possible if you have any doubts about your physical condition. There are certain diseases which may prohibit becoming a pilot, for example diabetes. On a more subtle level, decent physical coordination and responses are crucial in the learning process as well as achieving competency and safety as a pilot. Are you the kind of person who can physically handle demands involving doing a number of things at the same time and in short order? If you are the "all thumbs," not only will the learning process probably be much slower, but you may face the fact that you can never really achieve needed competency.

8. Am I psychologically able to learn to fly?

When you deal with whether you can handle the psychological strains and stresses, you are treating a wide-ranging and yet very personal subject. It includes questioning yourself as to whether you're willing to tackle a new field, a really new field of academic study ranging from weather phenomena to aeronautics to navigation and throughout, accept the challenge of becoming knowledgeable in these fields. Flight instruction is a demanding course; it has to be for your safety and the safety of others. Beyond the demands of flight and ground instruction, will you be able to handle being on your own in a plane, shouldering responsibility for yourself and ultimately for loved ones, friends, and others? Are you willing to undergo a rigid testing procedure involving a written examination of your knowledge as well as a flight test in the air? The psychological strains and stresses of learning to fly should not be underestimated.

9. Do I have the support of those close to me?

Think about the wife, husband, sweetheart, parents, or others as the case may be. You need their support and, in most instances, need it vitally. It is fair to say that you can't ultimately "win" unless they are behind you; true, some will change their mind about your undertaking flight training but it is fair to say that the person who is going to best be able to handle the stresses and demands of learning to fly will have the support of those who are important to that person in undertaking the endeavor.

10. Do I realize the limitations of flying even after I get my Private Pilot's License?

The VFR (Visual Flight Rules) pilot (the average pilot) is limited to flying in really good weather. Given the vicissitudes of weather in the United States, you simply can't fly too far, too long, before weather changes so as to prevent further flying. You must not only realize and accept (there's real trouble if you don't) these limitations, but also the danger that job and family may generate a case of "gethomeitis"—and that can mean disaster.

Give some careful thought to your interest in learning to fly. If any of these ten questions yield answers which realistically cloud the possibilities, then at least you owe it to yourself to take a deeper look at the subject before taking that first lesson. Keep in mind that probably between 60 and 80 percent of those who undertake flight training never finish and get their private pilot's license. The most sensible answer for you may be not to undertake flight training or to delay it until some of these matters are changed or cleared up. On the other hand, reasonably positive responses to the question raised would seem to indicate that you are ready to undertake flight training because you have asked and answered some important questions about undertaking flight training.

Appendix II
Basic Steps

The aim of flight training and ground schooling is to obtain a license (certificate) from the Federal Aviation Administration (FAA) which allows you to pilot a private plane *with passengers* aboard. Actually, once you have soloed and have had your log book properly endorsed by your instructor, you can do all the flying you want, but you cannot carry any passengers; furthermore, aircraft renters will want to see that private pilot's license. Realistically, then, the aim is to obtain a private pilot license.

The steps and requirements are stated in the usual order; there can be variations, but this gives you an idea of what's in store.

1. Make the decision to undertake flight instruction. See appendix I.

2. Determine where you will obtain flight training and ground schooling. See appendix III.

3. Make arrangements for your training—in terms of time to be alloted and money to cover costs—but see medical requirements under the next step first!

4. Obtain a Student Pilot Certificate and current Medical Certificate. You must have these certificates before you may fly solo. While there are different classes of medical certificates, all you need for your solo flying and private pilot's license is what is known as a Third Class Medical Certificate which is valid for twenty-four months. In order to get these certificates (these are back-to-back on the same piece of paper):

    A. Get a list of doctors who have been authorized by the FAA to conduct aviation medical physical examinations. Your flight school, instructor, or airport manager can tell you what doctors have been so designated.

    B. Take the physical. It will cost between $25 and $40 in most instances and takes about thirty minutes. The minimums include the following:

- 20/30 or better in the weakest eye when wearing glasses.
- no serious trouble with your eyes.
- be able to distinguish red, green, and white.
- be able to hear a whispered message from a distance of three feet.
- no internal ear trouble.
- normal breathing through your nose and throat.
- normal nervous system and sense of balance.
- no organic or functional disease which could interfere with flying. A heart analysis (electrocardiograph) is necessary only when the doctor believes there is a heart or blood circulatory condition.

If you feel you may have some deficiency or medical problem that might cause a problem, contact your local FAA district office or the FAA regional doctor; they will help you determine whether your condition, medication, or such will prevent you from obtaining a Medical Certificate.

C. The doctor who gives you your physical can also issue the Student Pilot Certificate which appears on the other side of the Medical Certificate is also valid for 24 months, which is more than enough time to get your Private Pilot's Certificate (or license). Even after you get your Private Pilot's Certificate, the Medical Certificate on the reverse side of that Student's Certificate must be kept in your possession when you fly because it remains your required medical credential.

5. Obtain a Restricted Radiotelephone Operator Permit. You need this permit in order to legally operate the 2-way radio in the airplane. Your flight school, flight instructor, or airport manager usually has a supply of application forms or you can write to the Federal Communications Commission (FCC), Gettysburg, Pennsylvania, 17325, for a form. There is currently no cost for this permit, but you are required to have one.

6. The following are the *minimum* experience requirements as set by the FAA in its regulations (Federal Aviation Regulation—FAR).

Non-Approved School (FAR part 61):
- Total Pilot Time—40 hours.
- Flight Instruction—20 hours, to include at least 3 hours of cross-country flying and 3 hours of night flying.
- Solo Practice—20 hours, to include at least 10 hours of solo cross-country flying.

Approved School (FAR part 141):
- Total Pilot Time—35 hours of which up to 5 hours of instruction in a suitable pilot ground trainer (simulator) may be counted.
- Flight Instruction—20 hours, to include at least 2 hours of cross-country flying and instruction in night flying.
- Solo Practice—15 hours, to include at least 10 hours of solo cross-country flying.

A. These are bare minimums; the average student has about 66 hours before taking the flight test.
B. See appendix III for explanation of non-approved and approved

schools; note that the approved school has just slightly lower minimums.

C. The night-flying requirement must be met only if you intend to fly (take off or land) one hour after sunset or one hour before sunrise. Otherwise, your license will be restricted to no night flying. You can later get the training, pass a flight test, and get the restriction removed.

D. The hours of flying experience are recorded in a flying logbook that you will keep and maintain; your instructor will sign the logbook after a flight, certifying the instruction you receive. You will log and certify your own solo time.

E. Cross-country flying is flight in excess of 50 nautical miles from the point of departure.

7. You must pass a written examination given by the FAA called the FAA Private Pilot Written Examination. Any training which prepares you for this exam and gives you needed aeronautical knowledge for safe, competent piloting is called ground school. This schooling is usually undertaken concurrently with flight training, though it can be completed before you ever step into an airplane. You must, however, pass the written test *before* you are qualified to take the Private Pilot Flight Test as it is called. See appendix III for further information on ground school training.

Ground schooling includes education and training in the following areas:

• The Theory of Flight
• Aircraft Powerplants
• Flight & Safety Practices
• Air Traffic Control
• Radio Operation and Navigation by Radio
• Meteorology and Weather Recognition
• Weather Maps, Reports & Forecast
• Cross-Country Navigation
• Flight Instruments
• FAA Rules & Regulations

There are a number of other study areas, in fact too many for this brief outline. You learn to use a flight calculator, prepare for a flight, and then "fly" a cross-country flight using maps, your calculator, and a flight plotter. You should count on spending many hours in your studies.

There is no charge for taking the examination. It consists of 60 multiple-choice questions and you must achieve a minimum passing grade of 70 percent; thus you must correctly answer 42 questions. Currently, the FAA publishes a booklet with 600 questions from which your 60 will be selected. The 60 questions cover a spread of topics you are expected to know.

The test usually takes 2½ hours to complete; you have 4 hours allowed.

The FAA recommends that a student pilot take the written examination after completing the first solo cross-country flight, because by that point in his flight training, he has gained operational experience that will help make the test more meaningful and easier. You can retake the test if you fail. Results are sent in about seven to ten days. You must present the Airman Written

Test Report (your results) showing a passing score to the person administering the Private Pilot Flight Test.

8. Take and pass the Private Pilot Flight Test. This is your final step; you have undergone the ground school training, you have logged the required number and kind of flight hours, you have a plane (rented or provided by your flight school) and you have passed the written examination.

To summarize,

To be eligible for a Private Pilot Flight Test, you must:

• Be at least 17 years of age.
• Have a medical certificate issued within the previous 24 months.
• Have a Student Pilot Certificate issued less than 24 months ago with solo and cross-country indorsements by your flight instructor.
• Have an Airman Written Test Report proving you have passed the FAA Private Pilot Written Test within the preceding 24 months.
• Completed the training required by FAR part 61 or 141, as applicable.
• For FAR part 61, you must show a logbook with at least the minimum flying experience logged. For FAR part 141, you must show a graduation certificate from an "approved" school which assures the flight examiner that all training and prerequisites have been met.
• Have a properly completed application for an airman certificate and/or rating signed on the reverse side by your flight instructor within the previous 60 days.
• Have a Radiotelephone Operator Permit.
• Have reasonable weather for the flight test. This is usually considered to be a ceiling of 3,000 feet or higher, 3 miles visibility or greater, and winds of less than 20 knots.

The Flight Test is given by an FAA flight examiner or a certified flight instructor who has been designated to give this test. FAA examiners do not charge for the flight test (but you are paying for the rental of the airplane); designated flight instructors do charge approximately $25.00 to $35.00, whether you pass or not.

The flight test takes about half a day and involves what is termed an "oral" part, testing your aeronautical knowledge, preflight inspection operation procedures, and knowlege of the aircraft's performance and limitation; you then prepare a flight plan for the proposed cross-country flight, gathering weather information and plotting the course. Once airborne, you are put through the paces—performing flight maneuvers, flying at slow airspeeds, dealing with stalls, performing good landings at airports along the way, handling emergency procedures, correctly navigating the plane to directed places, using radio communications and radio navigation aids effectively and generally exhibiting the experience and judgment necessary for your own safety as well as the safety of others.

If you pass, you immediately get a Temporary Airman Certificate, good for flying until your permanent certificate is sent to you. If you fail, you get a Notice of Disapproval of Application telling you what procedures or maneuvers you failed. You then must take additional training, and, after certification by your instructor, you are retested on those portions you failed.

Thereafter, you must keep your proficiency up and undergo a biennial flight review every two years, to be certain you have maintained your proficiencies.

Appendix III
Where to Go to Learn to Fly

The decision to learn to fly is a serious one; the result can be the joyful opening of a new dimension for life. This section is designed to help you find the right flight training, once you have made the decision to learn to fly.

FAA Approved Schools

By having certain facilities and meeting certain qualifications, a school can be deemed "approved." These are known as Part 141 Schools since they are qualified under those FAA regulations and requirements.

These schools are almost invariably residence institutions; you are a "full time" student living on the campus or in the vicinity of the school. Typically, such schools offer training toward many levels of flight certification and generally are aimed at those interested in aviation as a career.

They offer intensive, accelerated training in flight instruction and ground school. Some weeks must be spent in preparation for the Private Pilot Flight Test and Written Examination, but the investment of time and money is substantial. If you are in a hurry and want to really devote a full time effort, this route to your license should be considered.

Here is a list of some of these schools; drop them a line or call and they will send you their catalog. Tell them you are interested in accelerated pilot training:

Acme School of Aeronautics
Terminal Bldg. Meacham Field
Fort Worth, Texas 76106
817-626-2444

American Flyers, Inc.
P.O. Box 3241-C
Ardmore, Oklahoma 73401
405-389-5471

Sparton School of Aeronautics
8820 E. Pine St.
Tulsa, Oklahoma 74151
918-836-6886

The School of Aeronautics
Florida Institute of Technology
P.O. Drawer 1839
Melbourne, Florida 32501
305-727-7544

Ross School of Aviation
Riverside Airport
Tulsa, Oklahoma 74107
918-299-5056

Burnside-Ott Aviation Center
Bldg. 106
Opa Locka Airport
Miami, Florida 33054
800-327-5862

Sowell Aviation Co., Inc.
P.O. Box 1490
Panama City, Florida 32401
904-785-4325

Emery Aviation, Inc.
Rt. 4, Box 173
Greeley, Colorado 80631
303-352-8424

FAA Non-Approved Schools

Don't let the "non-approved" label deceive you. There are many schools or flight training centers that simply do not choose to be of the "approved" type, but they offer quality flight and ground school instruction by certified flight instructors and thus are fully qualified to train you to fly.

Many of the flight training centers or pilot training centers are associated with one of the major aircraft manufacturers. These flight training centers or flight schools are where most people learn to fly. The manufacturers have prepared flight courses of instruction integrating flight training (in their airplane, of course) with ground school instruction.

Generally, you fly by the hour for the rental of the plane plus an hourly fee for instruction. You will usually have to invest up to $100.00 for basic text and other materials when you undertake the pilot training course; this investment makes you a "student" participant of the course and entitles you to the use of the audio-visual materials and other ground school training offered by that course.

There are strong similarities between the courses or schools offered by major light-plane manufacturers through their authorized dealers. All integrate flight instruction on an hourly basis with ground school training, the latter using audio-visual materials (tapes, film strips). Usually there is no binding contract commiting you to finish or to pay beyond the initial enrollment fee (where you get the text materials, computer, workbook, log book, etc.) and for the flight training where you pay for the rental of an aircraft and hourly time for instruction.

Rental costs for the usual trainer-type plane vary considerably throughout the country. Generally, however, the average will be about $30.00 per hour for the plane wet (with gas and oil) and instructor's time. The national average for persons undertaking flight training runs about 66 hours.

Ground schooling consists of home study of a text, doing workbook problems, and using audio-visual materials at home or at the flight training center. Quizzes are given at intervals and a final examination. Upon passing, the school can issue a "diploma" entitling you to take the written FAA examination. The school will also prepare you for the flight test and help you make arrangements for it through the nearest FAA office.

To find the places (always at local airports) where you can get this training, look in the Yellow Pages under Aircraft Schools. By all means, check out more than one of these courses. Most offer an "introductory ride" as a promotion for a nominal cost; usually $10.00. Try a high-wing ride as well as one of the low-wing aircraft and see how you feel about each as well as evaluating the type of operation being run. Don't be afraid to ask questions such as:

• How many trainer-type aircraft do you have available?
• How many Certified Flight Instructors do you have full time? How many part time?
• What are your charges for flight training?
• How do you organize your ground school training?
• Do you have a hangar for your trainer-type planes so that training can continue in the winter?

If you can only take lessons on the weekends, be certain time with instructors and planes will be available to you. Check out the financial arrangements; do they accept charge cards; are financing plans available?

The following is a list of manufacturers that offer a pilot training course at authorized or franchised dealers:

Cessna Aircraft Company
P.O. Box 1521
Wichita, Kansas 67201

Piper Aircraft Corporation
Lock Haven, Pennsylvania

Beech Aircraft Corporation
Wichita, Kansas 67201

Gulfstream-American
(formerly Grumman American Aviation Corporation)
P.O. Box 2206
Savannah, Georgia 31402

Write to any of these manufacturers and ask them for a list of their pilot training centers and a description of their course in private plane piloting.

The General Aviation Manufacturer's Association, to which all of the major single-engine aircraft manufacturers belong is running a contest, giving away one airplane every six months, the winner drawn at random from new pilots licensed between January 1, 1977 and December 21, 1979. They will send you information on the participating flight schools in your area. Contact:

GAMA
P.O. Box 3990
Peoria, Illinois 61614
800-447-4700
(in Illinois: 800-322-4440).

More information on flight and ground school instruction can be obtained by writing the

Department of Transportation
Distribution Unit, TAD-443.1
Washington, D.C. 20590

And ask for

FAA Advisory Circular
AC 140-2K, "List of Certified
Pilot, Flight & Ground Schools"

Also request a list of non-approved schools in your area. Remember, that any flight and ground school listed (and thus regulated) by the FCC can give flight training and ground school instruction preparatory for the written and flight test requirements for a license. Also keep in mind that *no one but a certified instructor* can give you flight training. This is not an area where a pilot "friend" can train you. You will, however, find free-lance certified flight instructors at almost every airport (including thousands of small airports you may have never heard of though they are in your area); they are not attached to any "school," course, or aircraft dealer, but they can train you, supervise your ground school instruction, and in fact give you the necessary "diploma" qualifying you to take the written examination, as well as preparing you for the flight test.

Ground School Instruction

As you have noted above, ground school instruction can be gained in a number of ways including:

1. As part of residence training at an approved school.
2. As part of your course of instruction at a non-approved school, such as run by aircraft dealers offering the training organized by an aircraft manufacturer;

In addition, ground school instruction can be gained by:

3. The supervised study had with a free-lance certified flight instructor.
4. Attending a ground school offered by a high school or college at night, in many areas.

5. Businesses that offer weekend ground school courses on an accelerated basis (the entire course in two or three days); you will find advertisements in any of a number of magazines devoted to flying. — See recommended reading.
6. Through home study courses, including at least one on tape.

There exists a great deal of flexibility here, and you may find that you will try more than one consecutively or simultaneously in order to achieve the results that are best for you.

Glossary

Ailerons – The moving portion of the wing that moves the plane around its roll or longitudinal axis. Turning the wheel from one side to the other moves the ailerons on the wings. They roll or bank the plane.

Airspeed – The speed of an airplane as it moves through the air, independent of distance covered on the ground.

Altimeter – An instrument that uses barometric pressure to indicate altitude of the plane.

Attitude – The relationship of the nose of the airplane to the horizon.

Automatic Direction Finder – Radio navigation instrument.

Carburetor heat – A device used to circulate hot air around the carburetor to prevent ice from forming.

Ceiling – The height above the earth's surface to the lowest layer of clouds. A weather term; the cloud cover over the flying site, measured in feet from the ground to the bottom of a defined layer of clouds.

Certified Flight Instructor (CFI) – A pilot-teacher, certified or licensed as such by the FAA.

Cross-country flying – A term used to describe the requirements of dual and solo flights during training involving flights of over 50 nautical miles from takeoff; also commonly used to describe flying anywhere but locally.

Dead (deduced) reckoning – Involves calculating the elapsed time from point to point under anticipated conditions of flight, including wind, groundspeed, etc.

ETA – Estimated time of arrival.

Elevator – The horizontal airplane tailpiece that moves the plane in its pitch or lateral axis; it is moved by forward or backward pressure on the wheel or yoke; raising or lowering the nose of the plane.

Federal Aeronautics Administration (FAA) – The federal agency basically charged with regulating airplane manufacturing and all phases of flying including pilot instruction.

Federal Aviation Rules (FARs) – The rules and regulations concerning airplanes and all phases of flying promulgated and supervised by the FAA.

Flaps – Sections of the wing lowered by the pilot to change the airfoil configuration of the wing; used in landing for proper angle of descent.

Flight plan – A form filled out by the pilot prior to takeoff and then communicated to a flight service station; it contains essential information concerning the plane and its planned destination.

Flight service station – System of offices maintained by the government to give in-flight assistance of a variety of sorts.

Flight watch – An in-flight weather advisory service; by radio a pilot can obtain updated weather reports as he flies.

Ground control – At larger airports, a separate control facility directing airplanes on the ground as they move around the airport.

Ground school – The term describing all of the aeronautical studies necessary for flying and preparing for the written and oral pilot license examination.

Hood work – The training procedure in which the student pilot wears a hood over his forehead so that he can't see anything but the flight instruments.

Instrument Flight Rules (IFR) – Those rules regulating flight by a pilot who is essentially depending on aircraft instruments for flying; see VFR.

Instrument-rated pilots – An advanced rating for the private pilot involving training in flying the airplane using only instruments in the cockpit, allowing flight in weather conditions in which the non-instrument-rated pilot could not fly.

Landing pattern – The flight path flown by a pilot in order to properly prepare for landing at any airport.

Magneto – Device that supplies electrical power for airplanes ignition system.

Minimum controllable airspeed – Flying at the slowest speed possible without losing altitude or stalling the plane.

Pilotage – Flying from a seen point to the next point to be observed.

Private Pilot's Certificate – The certificate or license granted by the FAA after completion of both the written examination and flight test; with it you can carry passengers as you fly.

Private pilots flight test – FAA administered or supervised test; includes an oral examination and a test of flying ability in a plane with an examiner.

Private pilot's written examination – FAA administered test covering all phases of required aeronautical knowledge.

Restricted Radiotelephone Operator Permit – Permit in order to legally operate the 2-way radio in the airplane.

Rudder – The vertical airplane tailpiece, moved by foot pedals that move the plane around its yaw or horizontal axis; moving the nose to the left or right.

Run-up procedure – The use of a checklist by the pilot to test the airplane just prior to takeoff to make certain everything is functioning properly.

Sectional chart – A special aeronautical map of a section of ground terrain.

Slow flight – Flying the plane at minimum controllable airspeeds; a skill vital to proper landing technique.

Solo – Flying the airplane without assistance from anyone else.

Stall – Placing the airplane in an attitude in which the angle of attack of the wing is so excessive that lift is destroyed and the plane noses down.

Student Pilot Certificate & Medical Certificate-A document with the Student Pilot Certificate (license) on one side and Medical Certificate on the other issued by an FAA designated physician and necessary before soloing a plane.

Tail-dragger airplane-Any aircraft where the tail rides on a wheel or skid on the ground. — (See tricycle-gear airplane.)

Touch-and-go's-The basic landing practice procedure in which the plane is landed to the point of touching the runway and then brought up and around a landing pattern for another "touch-and-go" around.

Transponder-A radio receiver-transmitter device that helps identify your airplane on a radar screen by sending out a discreet signal.

Tricycle-gear Airplane-Any aircraft where the nose or front of the aircraft rests or rides on a wheel. — (See tail dragger airplane.)

Trimming out-Adjusting the controls of an airplane so as to achieve stable straight and level flight.

VOR-A transmitting station continually broadcasting radio signals that can be used by a pilot to determine where he is and to assist in flying from one place to another.

Victor airway-Specific, frequently traveled line of radio navigation; tied into the VOR navigation system.

Visual Flight Rules (VFR)-Those rules regulating flight by a pilot who is essentially flying by reference to what he sees as against flying by instruments alone; see IFR.

Weather briefing-Required of every pilot before any cross-country or extended flight; obtained from government weather sources and includes local, enroute, and destination weather.

Index

Acknowledgments

Those who have helped me along the way, both in terms of learning to fly and writing this book, deserve my thanks. They include Herb Merel, John Bishop, and Sid Axelrod on the radio controlled modeling aspect of flying, along with Phil Kraft and Marty Barry whose world bridges both radio-controlled and full-scale flying.

An acknowledgment of gratitude that is well deserved goes to Tom Murray, my flight instructor and Rick Bodee, Chief Pilot, as well as all of the people at T.K. Aviation, where I took my flight training. My special thanks to Ed Arnolds and Mike McKinley for their encouragement as well as being flying buddies.

I owe a special debt of gratitude to my friend and former history colleague, Dr. Duke Frederick, for reading and correcting the manuscript.

Recommended Reading

Magazines on flying

Private Pilot

Plane & Pilot

Flying

Air Progress

The Aviation Consumer
Box 972
Farmingdale, New York 11737

Aviation Monthly
United Media International
306 Dartmouth Street
Boston, Massachusetts, 02116

FAA General Aviation News
Superintendent of Documents
Government Printing Office
Washington, D.C. 20402

Books about flying

Langewiesche, W., *Stick and Rudder*, McGraw Hill.

Fillingham, Paul, *Basic Guide to Flying*, Hawthorne.

Back to Basics, Editors of *Flying* Magazine

Pilot Error, Editors of *Flying* Magazine

Best of Flying, Editors of *Flying* Magazine, Van Nostrand Reinhold.

Burnham F., *Cleared to Land!* The FAA Story, Aero Publications.

Carmichael, A & D, *From White Knuckles to Cockpit Cool*, Aero Publ.

Gilbert, *The Flier's World*, Random House

Hoyt, *As the Pro Flies*, McGraw-Hill Book Co.

Taylor, *Understanding Flying*, Delacorte Press